About Island Press

Since 1984, the nonprofit organization Island Press has been stimulating, shaping, and communicating ideas that are essential for solving environmental problems worldwide. With more than 1,000 titles in print and some 30 new releases each year, we are the nation's leading publisher on environmental issues. We identify innovative thinkers and emerging trends in the environmental field. We work with world-renowned experts and authors to develop cross-disciplinary solutions to environmental challenges.

Island Press designs and executes educational campaigns, in conjunction with our authors, to communicate their critical messages in print, in person, and online using the latest technologies, innovative programs, and the media. Our goal is to reach targeted audiences—scientists, policy makers, environmental advocates, urban planners, the media, and concerned citizens—with information that can be used to create the framework for long-term ecological health and human well-being.

Island Press gratefully acknowledges major support from The Bobolink Foundation, Caldera Foundation, The Curtis and Edith Munson Foundation, The Forrest C. and Frances H. Lattner Foundation, The JPB Foundation, The Kresge Foundation, The Summit Charitable Foundation, Inc., and many other generous organizations and individuals.

The opinions expressed in this book are those of the author(s) and do not necessarily reflect the views of our supporters.

Don't Be Such a Scientist

SECOND EDITION

Don't Be Such a Scientist

Talking Substance in an Age of Style

SECOND EDITION

By Randy Olson

Washington | Covelo | London

ISLAND PRESS is a trademark of the Center for Resource Economics.

Library of Congress Control Number: 2017958880

All Island Press books are printed on environmentally responsible materials.

Manufactured in the United States of America
10 9 8 7 6

Keywords: *An Inconvenient Truth*, anti-science movement, academia, climate change denial, Michael Crichton, documentary, evolution, film, *Flock of Dodos*, Al Gore, Hollywood, improv, listening, Carl Sagan, science communication, messaging, narrative, National Academy of Sciences, scientists, Story Circles, storytelling

This second edition is dedicated to my mother, Muffy Moose,
who always said, "You gotta shake 'em up!"

Contents

INTRODUCTION TO THE SECOND EDITION   1

CHAPTER 1   Don't Be So Cerebral   19

CHAPTER 2   Don't Be So Literal-Minded   53

CHAPTER 3   Don't Be Such a Poor Storyteller   89

CHAPTER 4   Don't Be So Unlikeable   127

CHAPTER 5   Don't Be Such a Poor Listener   157

CHAPTER 6   Be the Voice of Science!   181

APPENDIX: FILMMAKING FOR SCIENTISTS   213

ACKNOWLEDGMENTS   219

NOTES   221

INDEX   227

Introduction to the Second Edition

"You think too much! You motherf***ing think too much! You're nothing but an arrogant, pointy-headed intellectual—I want you out of my classroom and off the premises in five minutes or I'm calling the police and having you arrested for trespassing. And I'm not f***ing joking, you a**hole."

Well. That was the opening to the introduction of the first edition of this book.

It's been nine years since it came out, and twenty-three years since the crazy acting teacher (as I've affectionately come to call her over the years) screamed those words at me. I've had a lot of time to process what she meant. One thing is certain—it's still the most important moment of my entire professional life.

I was an innocent thirty-eight-year-old aspiring filmmaker who had just resigned from my tenured professorship of marine biology at the University of

1

New Hampshire. I had moved to Hollywood and entered her class on the advice of a Hollywood veteran. But I wasn't ready for what happened. When I look back today—now in my sixties—there is "before CAT" (crazy acting teacher) and "after CAT." It's that simple.

I will always hate her for what she did to me. She singled me out. It was the very first night of the two-year program. She had warned us we were not allowed to ask questions in class—only listen. But I was an academic—I had been trained to question everything. When I finally couldn't help myself, she let me ask one skeptical question. And then she lost control and unleashed a tirade about having worked with "your type" that ended with the red-faced, vein-popping climax above.

I had never been subjected to anything like it in my life; I left the class in shock. We eventually worked it out and I went the full distance with the course. But it would take two decades to process why she didn't want me there. Furthermore, I'm still learning from it.

I will always hate the teacher for that one humiliating night, yet I will also love her for opening my mind to a whole different view of the world. Life is meant to be a journey of learning. And now, for me, the journey has led me to one central focus: developing a deeper understanding of narrative. It lies at the core of not just communication but the human soul itself.

You might scoff at the flakiness of this last sentence, but try reading the preface Christopher Vogler wrote in 2007 to the second edition of his landmark 1992 book, *The Writer's Journey*. The book is arguably the most important resource on storytelling for Hollywood screenwriters. It's a detailed examination of "the Hero's Journey," which, as mythologist Joseph Campbell pointed out, is the central template for storytelling around the world.

Vogler says, "I came looking for the design principles of storytelling, but on the road I found something more: a set of principles for living. I came to believe that the Hero's Journey is nothing less than a handbook for life, a complete instruction manual for the art of being human."

He's talking about narrative and he's talking about being human, which is what I nowadays want to talk about more than anything else. So here we go with a new introduction, written in the summer of 2017.

Dog Ears

Nothing warms the heart of a writer more than looking up to see a dog-eared copy of your first book in the hands of someone waiting in line at a book signing. I've had plenty of those moments since the publication of *Don't Be Such a Scientist* in 2009. The best was at a Washington, DC, conference for the National Park Service.

A woman in her mid-thirties, accompanied by her husband, stepped up to the table where I was signing copies of a later book. She nervously held a tattered copy of *Don't Be Such a Scientist* and asked me to sign it. With a certain degree of shy awkwardness, and mostly looking away, she told me the book had changed her life.

She said that a few years back she had been working on a Ph.D. in conservation biology. As much as she loved the animals she was studying—bats—she just didn't feel she was meant for a career as a research scientist. A friend gave her a copy of *Don't Be Such a Scientist*.

The book made her appreciate how much she loves doing science, but it also made her realize that what she loves even more is telling others *about* science. After reading almost the entire book in one sitting, she made a major life decision: she quit work on her doctorate and shifted to a master's degree in communication. She now works as head of communication for a bat conservation organization and is thoroughly enjoying her life. I could sense her appreciation in the nervousness of her voice. After she recounted a couple of her favorite parts of the book, her final words to me were, "Thank you for showing the human side of science."

Those words rang deep. If I had to pick one overarching message I've tried to convey with my communications efforts—this book, my other books, my films over the past thirty years, and all the teaching I've done—it's been the need to show that this wonderful thing called science is really ultimately about humans. That is an endlessly difficult challenge.

There is a force at the core of science that drives it in a nonhuman direction. You see it over the ages. From the creation of robots to the destruction of humanity through war technology, science left to its own devices drifts in a direction that is not good. It doesn't happen with evil intentions, only through a lack of self-awareness. Science brings so much benefit to humanity, but it also needs voices to constantly pull it back from this nonhuman direction (and now more than ever, given the rapid onset of AI, artificial intelligence).

In writing this book, I didn't set out with any more of an agenda than to identify the major problems I saw in the communication of science, both within science and to the public. But once it was completed, I could see a basic theme that underpins all I had to say and that continues to be my core message. The overarching message I had back then and still have today is that we must all work constantly to Make Science Human.

Solar-Powered Scorn

When *Don't Be Such a Scientist* came out in 2009, it received plenty of rebuke. Actually, before it even came out, the title and cover artwork were immediately met with displeasure by a number of science bloggers (many of whom were the same people I complained about in the book). They looked at the title and said, "We don't need such a negative-sounding book about science."

I was familiar with the complaint from my previous works, such as my documentary feature film *Flock of Dodos: The Evolution–Intelligent Design Circus*. A group of scientists at the Smithsonian Institution—many of whom I knew from graduate school—managed to kill a scheduled screening of the movie there for similar reasons: they felt that "the movie doesn't paint a flattering portrait of scientists." To which my reply has always been: I thought we were trained as scientists to present the truth, not flattery.

For this book, one of the great joys was the day in the summer of 2010 I spent being hosted by the Centers for Disease Control and Prevention in Atlanta. It was one of the most challenging and deeply rewarding experiences of my professional life. But halfway through the day, one of the organizers pulled me aside and quietly told me that the week before, a group of scientists had tried to veto my visit. They had been under the impression that the book was an anti-science screed. None of them had read it; they had just failed to notice the word "such" in the title. They thought the message of the book was "Don't be a scientist!"

I'm used to the blowback. And the fact is, the book does spend quite a bit of time critiquing scientists for how they go wrong with communication. But it's not entirely criticism. It ends on an enthusiastically positive note, as the last chapter is titled "Be the Voice of Science!"

The price I have paid for not flattering scientists has included ostracism, countless snide things said (my favorite being the one-star Amazon reviewer who said he hated this book with "the fiery passion of a thousand suns"), and snubs from science organizations, irked at not being worshipped. Yet the irony is, as a friend pointed out, on any given day if you search for "science communication" on Amazon, you'll more than likely find my 2015 book appearing in the number one slot. Which means plenty of people are at least interested in my message of making science human.

At this point, I'm used to it. The stuffed shirts can't stand the simplicity and candor of my approach. Yet I know that with at least some folks—like that woman in line at the book signing—the message has been heard.

Time to Get Back to Work

I wish I could say I feel like celebrating, but the fact is there is much more work to be done.

In the new material in this edition of *Don't Be Such a Scientist*, you will learn about what I now feel is the central problem for the communication of science. In 2009, I felt it was storytelling. This is evidenced by the longest chapter being the third one, "Don't Be Such a Poor Storyteller."

The chapter on storytelling still offers important lessons for scientists. However, I've moved a little deeper into the challenge. I've gotten smarter and now want to refine it further. It's not about storytelling—it's about narrative. They're not the same. Here's the difference.

Storytelling consists of three parts—the non-narrative phase (which Joseph Campbell called "the ordinary world"), where a story begins; the narrative phase (which Campbell called "the special world"), where the journey takes place in the search for a solution to a problem; and then back to the non-narrative phase, where all is back to normal.

The narrative phase starts when the problem is identified (in a murder mystery, a dead body is found) and ends when the problem is solved ("who dunnit" is revealed). On either side of the narrative part are the non-narrative parts—the setup in the beginning and the synthesis at the end (the long-winded speech from the hero, after he slays the bad guy, where he explains the moral of the story).

It all sounds very simple, but narrative turns out to be infinitely complex and can never be totally mastered. Yet even though it is infinitely complex, it has a simple core that can be captured by three letters—ABT.

They stand for the "And, But, Therefore" template I have devised over the past six years. I adapted it from people in Hollywood and, ultimately, from the 1700s philosopher Georg Wilhelm Friedrich Hegel as well as from the Greeks. The ABT is the spine of the Hero's Journey, which Vogler describes as the handbook of life. The hero lives in a world of peace AND contentment, BUT then a problem arises; THEREFORE, the hero sets out on a journey to solve the problem. Yes, it is that simple at the core.

Although I wrote chapter 3 before I created the ABT framework, it is really all about the ABT structure versus the non-narrative AAA (And, And, And) structure; I just didn't yet have the terminology to talk about it concisely. Now I do.

The new content for the second edition comes in three parts. First, there is this new introduction. Then there is a new chapter, "Don't Be Such a Poor Listener." Finally, each of the five original chapters is followed by a brief set of updating comments, each time titled, in true ABT fashion, "Therefore . . ."

So here we go.

We Have More Work to Do: There's an Anti-scientist in the White House

We have a potentially nightmarish problem at hand. Not only is there an anti-scientist in the White House (who might be gone by the time you read this), but things have turned so strange that the attacks on science—which seemed to be abating in 2009—are well and thriving. Since the publication of the first edition, things have gotten worse rather than better for science. There is now widespread agreement that the entire profession is under attack.

It is time for a second edition of this book. Not only is the mission of improving science communication unfinished; we're still in the middle of the journey. More specifically, we're at what story experts generally refer to as "the darkest hour." This is the point where, according to Matthew Winkler's brilliant TED-Ed video "What Makes a Hero?" (which I've shown a thousand times in my Story Circles Narrative Training program), the hero "faces death, and possibly even dies. . . ."

"The hero" in our case refers to all of us who respect science and care about the environment. We're on a mission. We have a problem. We're in the middle of a narrative (defined, as I said earlier, as the series of events that occur in the search for a solution to a problem).

Our problem is to someday create the "science-literate world" that Chris Mooney and Sheril Kirshenbaum dreamed of in 2009 with their book *Unscientific America*. We're going to prevail eventually, but right now, things are a little dark.

An anti-scientist has captured the White House and, at least as of this writing, remains in power. I'm sure he wouldn't call himself an anti-scientist, but the fact is he has so little commitment to science that he's willing to ridicule climate science, withdraw the United States from the Paris climate accord, and appoint a creationist (Jerry Falwell's son!) to his task force on reforming higher education. Those acts say enough. No true supporter of science would consider such things.

The recoil to having elected this enemy of science manifested in the science world in the spring of 2017 as 100,000 people came together in Washington, DC, and elsewhere in a March for Science. To me, it was the brightest, most hopeful event for the future I've ever seen. It also vaguely points in the direction science needs to head—a direction that perfectly matches the "don't be such a scientist" message of this book. So let's take a look at what happened in a little more detail.

And . . . the Problem Came to Life with the March for Science

Spontaneity. Not a common trait for the science world. I talk about it in detail in the first chapter of this book—how scientists lack it and how improv acting fosters it. Spontaneity is the antidote for the excessively cerebral.

Scientists are great if you let them control every single thing that is going to happen. You can see this in how experiments are run. They usually include these things called "controls." Does that term give you a little feel for how scientists feel about spontaneity? As soon as something unplanned crops up, watch out.

Such is the wonderful and inspiring story of the March for Science. Nobody at the top of the science world was involved with its inception.

In military terms, it was like low-level soldiers planning a mass rally by taking their plans to the top general, putting a gun to his head, and saying, "You're in favor of this, right?" That's essentially what happened with the March for Science when it came to the major science organizations—they were approached after the march was already planned. Many of them felt too rushed, too pressured, and declined to officially support it.

The idea for the march began with a discussion on the Internet site Reddit in January—just three months before the march itself took place. A few people were innocently talking about the Women's March, a month earlier, which had involved over 4 million participants. Someone mentioned the idea of doing the same thing for science. A small group agreed. They were just average folks—no heads of organizations. They organized a Facebook group, and to their surprise the membership began growing rapidly.

I spoke with Valorie Aquino, one of the three codirectors of the march. She said they had a conference call in which they mused over the 1,000 members the Facebook group now had. They ended their call with a clear plan of how to grow the group to about 3,000 within a month.

The next morning they awoke to the stunning news that the group had grown overnight to 10,000. Within a week it had passed 100,000. By the time I spoke with her, it was approaching a million. Clearly, they had struck a chord. But where did all the energy come from?

March for
Science
(the gut)

Figure I-1. The 2017 March for Science started spontaneously in the gut. It was a narrative mess, but . . . it made me feel something because it was human. Photo in the public domain; accessed via Wikimedia Commons.

My friend Aaron Huertas joined the communications team for the project. Initially we both felt the march needed a clear message, which means "a narrative"—a clear problem-solution dynamic. Now I see I was kind of wrong.

Ed Yong in the *Atlantic* pointed out the confusion by listing twenty-one messages that were being mentioned by organizers of the march. A number of articles were written arguing against the march itself. Many complained that the whole effort was politicizing science.

In the end, the organizers really weren't certain whether the event was meant to be a happy, fun science day for the family (like a science festival) or a more adult-oriented science version of the Women's March a few months earlier, which was filled with contempt for the newly elected president.

Putting the Mess into Messaging

The messaging ended up being a mess, but so what—turns out sometimes you don't need a message . . . yet. I remember arguing this in September 2011, when the Occupy Wall Street movement emerged. The protestors' narrative wasn't very clear during the first week they began generating mass attention. Lots of news pundits—including my longtime hero Chris Matthews of MSNBC's *Hardball*—criticized the movement, saying, "They don't know what they want."

But mass movements are almost never created by intellectuals with clearly thought-out plans. No, they generally arise from the masses, who are driven by the gut. Down the line, things move to the head—which is exactly what we eventually saw five years later as Senator Bernie Sanders began articulating a plan of action pursuing basically the same goals as had the disorganized youngsters who occupied Wall Street.

So, despite the rain that spring Saturday morning in Washington, DC, tens of thousands of happy, fun, enthusiastic people turned up and the event was a stunning success. People carried all kinds of wildly creative and inspiring signs. Speeches were given before the march. They weren't the sort of landmark speeches that have historically accompanied major protest events in Washington, DC. The speakers didn't really seem to know what to say because . . . there was no clear message. But again, so what?

At the end of the march, in front of the Capitol building, there was . . . nothing. Just a woman with a bullhorn telling everyone in the rain to visit the website and keep the effort going. And that still didn't matter. The event was all about the hour-and-a-half-long march itself and the sheer mass of humanity that was present, acting not like scientists but more like humans.

The crowd size was estimated at around 100,000. You could criticize it to pieces for not having a clear message if you wanted to impress your friends. Or you could just soak it all in and even feel some emotion about it. I went with the latter.

At just about the start time, I exited the Ronald Reagan Building, taking a break from a conservation event, and joined my old marine biologist buddy Bob Steneck of the University of Maine. We strolled down into the masses and marched from the start, near the Washington Monument. Within a few minutes we ran into Dr. Daniel Pauly, another old buddy and the famous fisheries biologist who coined the popular term "shifting baselines," which I talk about in the first chapter. About halfway up Constitution Avenue, Bob and I stepped out of the crowd to walk up the stairs at the IRS building and watch the river of humans flow past. It was downright breathtaking.

What hit me most was the age range of marchers. There were lots of families with kids. Some we talked to had no connection to the world of science—they were just there for the spirit of it.

Others carried amazing signs—like two little kids holding a sign saying, "This family has five scientists!" Another family pushed a young woman in a wheelchair with a sign saying "Thanking Science for Research on Multiple Sclerosis." That's the stuff that made me look down at the ground and choke back tears. It really hit you—there was a huge, supremely human element to the event.

But the obvious question was, why was the turnout for the March for Science, despite the pouring rain, so incredibly large and inspired? Here's my interpretation.

John Oliver for President of the National Academy of Sciences

The climate science community has managed to break new ground on the challenge of how to bore people. The effort was led by the Laurie David–produced movie *An Inconvenient Truth* in 2006, but countless others have followed in her plodding footsteps. This still needs to change, and it would if comedian John Oliver could be made head of the National Academy of Sciences.

Here's the problem. Effective communication is essential for science to work. My own drive to become a scientist was fueled by a number of older marine biologists I met as a student who were tremendous communicators because they were such great storytellers. They inspired me, not by their knowledge of the theories on how to communicate effectively but by their ability to tell stories that I found mesmerizing.

They include the late Bob Paine (to whom I dedicated the first edition of this book); my Shifting Baselines Ocean Media Project partner, Jeremy Jackson (whom I cite repeatedly); my undergraduate buddy Dianna Padilla (who wrote a blurb for this book); Paul Dayton (who coached me on how to survive scuba diving under the ice in Antarctica); and Bob Steneck (whom I just mentioned). These people are just a few of the great storytellers I met as a scientist.

But those in the climate community have not just bored people—they have lowered the bar to such a level of boredom that they have entirely redefined the word "communication." I've been railing against this for more than a decade. This spring, John Oliver brought the problem into focus with a single short bit on his HBO show *Last Week Tonight*.

It was in the wake of President Donald Trump's withdrawal of the United States from the Paris climate accord. Oliver said:

> This week the climate movement may have gotten a symbol to rally around. Because apparently it was never quite enough to motivate ourselves out of love for this large gassy orb [Earth photo], but maybe, just maybe we can now motivate ourselves to do something out of our loathing of *this* one [Trump photo].

He's talking about love versus loathing. "Love" is a word of agreement; "loathing" is a word of contradiction.

Now let's look at it in terms of narrative dynamics. The three fundamental forces of narrative are agreement, contradiction, and consequence. John Oliver was pointing out that for more than a decade the climate movement has attempted to succeed mostly through only one of those forces: agreement.

The climate community has endlessly presented the facts, which we can all agree on. But that didn't work. It led to the dreaded "And, And, And" structure, which is boring.

Take *An Inconvenient Truth*. It basically ended with Al Gore saying, "Now that we all agree global warming is happening. . . ." He tried to ignore climate change skeptics, giving them only about a minute of time by citing Naomi Oreskes' study of scientists who agree climate change is happening. He chose to ignore the element of contradiction, in the form of climate change skeptics, that was right before him.

That approach hasn't worked. What works is narrative, which draws on the forces of agreement, contradiction, and consequence and gets stronger with stronger sources of contradiction. Which is what Trump now provides.

If only the climate movement can take advantage of it.

So the challenge to science is clear—embrace the opposition, go to work defeating it, and dethrone the anti-science forces. Here's the problem.

But . . . "Scientists Don't Understand Media"
—Michael Crichton, 1999

The core problem is that we live in a media society, but scientists—who are the catalyzers of change in our technology-driven society—don't understand media. That's not just my obnoxious opinion; it's what science fiction author Michael Crichton said near the start of the profound, powerful, and prescient (three p's!) keynote address he delivered to the world's largest science organization, the American Association for the Advancement of Science, in 1999.

He had been asked to talk about "improving the media image of scientists." Instead, he said, who cares what the public thinks of *scientists*; what matters is what they think of *science*. To address this he laid out what, in retrospect, was pretty much a blueprint for how to deal with the anti-science movement, which, in the years since, has emerged and solidified.

Why should anyone have listened to Michael Crichton back then? You might automatically discount everything he had to say because of how he jumped the science track near the end of his life by becoming a loud climate change skeptic. I know a lot about this because, in 2007, I spent three months trading about forty e-mails with him as I was filming my movie *Sizzle: A Global Warming Comedy*.

He did indeed kind of lose his marbles. He warned me in his first e-mail, "I may be the most cynical person on this planet." But that was 2007. He hadn't yet reached that point in 1999. In fact, the extent to which his brilliant speech was ignored might have been part of what drove him in that direction.

Crichton was worth listening to because he probably knew more about the combined worlds of science and media than any other person, ever. He left a solid career in biomedical science to pursue his Hollywood interests. In 1994 he became one of only two people ever to have, simultaneously, the number one sales rank for a book (*Disclosure*), a movie (*Jurassic Park*), and a television show (*ER*). (The other was actor Tim Allen, who was not quite so scholarly.)

Carl Sagan never came close to that. Steven Spielberg never came close to that. The man knew media.

Let me add that, having had both an academic career and a media career, I find the comparison to be like the famous adage about death and comedy: "Dying is easy; comedy is hard." I can tell you, science is easy; media is hard.

Sorry, it's the truth. Scientists don't understand media, but they need to learn because we continue to be more and more consumed with it as a society.

In fact, the other important quote for today in that Crichton speech is this: "The information society will be dominated by those who are most skilled at manipulating the media." Guess who we elected president in 2016.

Understanding media is no longer a frivolous distraction; it is now the life's blood of who we are. But here's where there's cause for hope: media is narrative.

Therefore: Science Needs Narrative

Narrative is humanity's hope. I mentioned that we're currently in our darkest hour with the election of an anti-science president, but there is hope, and it lies not in science, per se, but in the humanities.

You're all welcome to go on being blind cheerleaders for science, and STEM, and STEAM, and whatever education-based solutions you feel the need to promote, but I don't see it. You're welcome to implore the public for more science funding and swear we're falling behind in the world. But I'm no longer comfortable with that message.

We've got plenty of science but not enough humanity.

Here is my message to the science world, if there's anyone out there listening (and this seems like a misguided hope, given the title of chapter 5). I am making an overarching plea to focus on the more human elements of life in general, and I am pointing out that there exists a tool with which you can achieve this.

The tool is called narrative. It is at the core of pretty much everything. Obviously it is at the core of storytelling. But it's also at the core of argumentation, logic, reason, and, yes, even the scientific method.

My forty-year journey from science to cinema has led me to what looks, to many, like a complete obsession with narrative. And it is, to a large extent.

I've just made a video with the American Association for the Advancement of Science that opens with the line "Narrative is everything." I firmly believe this.

But it's much deeper than that, and I make no excuse for this semireligious obsession. Let me go back to Christopher Vogler and add the paragraph that follows the quote I gave you at the beginning of this introduction. Look at what he says next, and look at how, at an intuitive level, he compares it to aspects of science:

> The Hero's Journey is not an invention, but an observation. It is a recognition of a beautiful design, a set of principles that govern the conduct of life and the world of storytelling the way physics and chemistry govern the physical world. It's difficult to avoid the sensation that the Hero's Journey exists somewhere, somehow as an eternal reality, a Platonic ideal form, a divine model. From this model, infinite and highly varied copies can be produced, each resonating with the essential spirit of the form.

All of you hardest of the hard-core scientists—including my old colleagues who felt certain I was having a midlife crisis in 1994 when I gave up my tenured professorship of marine biology to move to Hollywood and attend USC film school—please now listen.

If you put Vogler's two paragraphs together, you'll see that he's pointing to the basics of narrative as providing "a handbook to life." Don't you think every science student should be deeply acquainted with this handbook to life? Is that too much to ask? Don't you think there should be a section on narrative at the start of every introductory textbook in all the major fields of science?

Just a few weeks ago, I was sent a copy of the ninth edition of *Campbell Biology*. I flipped through the book. It's just the same old information, information, information. It's so stodgy. There's not one word about how science is half knowledge, half communication. The whole damn book is so hopelessly stuck in the sedimentary muds of the Oligocene.

And making it worse, I had lunch recently with my old buddy Joe Levine, coauthor of Miller and Levine's *Biology*, the most popular college biology textbook ever. He explained to me the harsh, cold constraints of textbook publishing. He said it's a zero-sum game—meaning if you add something about communication you're going to have to chop something informational, like how starfish reproduce. Tearing information out of the hands of science educators is about as likely as prying Charlton Heston's gun out of his cold, dead hands.

If I had my way, every introductory science book, from biology to physics to astrophysics, would open by mentioning how important communication is in today's world. It would also cite and promote the foundational book *"They Say / I Say"* by my buddy Jerry Graff and his wife, Cathy Birkenstein. Jerry was in my movie *Flock of Dodos*. This simple book, which presents the templates for use in argumentation, has sold nearly 2 million copies since its first publication, in 2006. Whenever I speak to mixed audiences of science and humanities students, I ask who has read it, and half of the hands go up—those of the humanities students.

Why should scientists go on being so oblivious of what is happening out in the human world? Every person in science should be required to read their book. And guess what—their title, *"They Say / I Say,"* is just the first two forces

Figure I-2. This is the sign I would have carried in the March for Science (had I been motivated enough to make the sign). Illustration by Kristin Black, www.etsy.com/shop/ProtestNation.

of narrative—agreement, contradiction. Same as the AB of the ABT. This stuff is universal. It's time to open everyone's eyes to it.

Oh, well. I suppose it's a long process bringing about change and making science human, but it needs to start somewhere.

Therefore, Make Science Human

This, then, is what Michael Crichton was indirectly talking about with that simple quote about scientists not understanding media. Media is narrative. Narrative is story. Story is the handbook to life. If only scientists would fully understand, grasp, and promote narrative as the central tenet of their profession, they would both understand media and make science human.

But the problem is that story—or, more specifically, narrative—is not something we learned analytically in childhood. Yes, when we were kids we told and listened to stories endlessly, but making use of this force in adulthood takes a lot more than what we still have from back then.

Just look at every major movie studio in Hollywood. They all have something called a Story Department. What I wish for the immediate future, and am certain will emerge by the distant future, is that every scientific institution in the world have a Narrative Department.

This would be at least one individual who is deeply trained in the core principles of narrative, knows how to analyze it, and knows how to fix it. That individual would be responsible for making sure that all communications released have strong narrative structure and, in so doing, would pull communications out of the trenches of boredom and confusion and up into the realm of the interesting, with which the general public can connect.

This does not mean inserting humor and emotion and sexual content into the serious work of scientists. Anyone who thinks that is what "narrative" means is as sadly misinformed as a lot of science communication theorists. It means understanding that narrative is about the basic problem-solution dynamic; that the brain is lazy; that achieving good narrative structure is an absolutely endless, energy-draining challenge; and that it needs more than an individual effort to make it happen.

If the science community understood narrative, it would understand anti-science attacks and be able to defend itself against them. It would be able to manage the societal implementation of the knowledge it generates, and it would be able to lead society, rather than being stuck taking orders.

This latter bit is nowhere more obvious and evident today than in the ongoing trend of there never being more than 1 or 2 genuine scientists among the 535 members of Congress. That pretty much says it all.

And that was, to some extent, what the March for Science was about. Not just rage over the election of a president who is openly hostile to science and academia, but also frustration with the lack of leadership for the science world. We need to have leaders for science who are bilingual in the languages of science and narrative. It's not too much to ask. It just needs to start with those textbooks, sending the message from the start that scientists need to be human.

CHAPTER 1

Don't Be So Cerebral

In 2000 Premiere *magazine ran an article about the making of the movie* The Perfect Storm. *The actor Mark "Marky Mark" Wahlberg talked about filming scenes off the coast of Massachusetts and told of glancing over his shoulder and spotting gray whales passing nearby. Even though it had been six years since I had resigned from my professorship, the scientist's eye never fades, and I couldn't help but be tripped up by that detail. I wrote a letter to the editor of the magazine explaining that those whales were either something other than gray whales (long since extinct in the Atlantic Ocean) or stunt doubles flown in from the Pacific Ocean. They published it. A couple of months later I ended up at a Hollywood party, spotted the issue of* Premiere *with my letter, proudly said to the group, "Hey, everybody, listen to this," and then proceeded to read my letter to the editor aloud. When I finished I looked up, beaming, but instead of applause I saw expressions of "Huh?" My best friend from film school, Jason Ensler, finally broke the tension by saying, "You know, the thing about Randy is, half the time he's like the coolest guy any of us know in all of Hollywood. But the other half of the time . . . he's a total dork."*

So we begin with the crazy acting teacher and some of the simple concepts she pounded into our heads night after night. There was one that emerged supreme seven years later, when I returned to working with academics. It is so simple and yet so powerful that I choose to start this first chapter with it. Most of what I have to say descends from this notion.

Here it is . . .

The Four Organs Theory of Connecting with the Mass Audience

When it comes to connecting with the entire *audience, you have four bodily organs that are important: your head, your heart, your gut, and your sex organs. The object is to move the process down out of your head, into your heart with sincerity, into your gut with humor, and, ideally, if you're sexy enough, into your lower organs with sex appeal.*

That's it. Others have heard me mention this in talks and put their own spin on it—talking about the chakras and "mind body spirit" and other sorts of New Agey gobbledygook. Also, there's vast work in the field of psychology exploring these sorts of dynamics. Carl Jung talked about personality types, and the Myers-Briggs Type Indicator, developed during World War II, explores this vertical axis of powers in the body. But, for our purposes, let's keep it simple and free of psychobabble. If you've had lots of classes in psychology, you may find this annoyingly simplistic. If not, I hope you'll find it as useful as I have.

It's about the difference between having your driving force be your head and having it be your sex organs. There *is* a difference.

Let's begin by considering each of the four organs.

The *head* is the home for brainiacs. It is characterized (ideally) by large amounts of logic and analysis. When you're trying to reason your way out of something, that's all happening in your head. Things in the head tend to be more rational, more "thought out," and thus less contradictory. Academics live their lives in their heads, even if it results in their sitting at their desks and staring at the wall all day, as I used to at times. "Think before you act" are the words they live by. When they ask, "Are you sure you've thought this through?," they are reflecting a sacrosanct hallmark of their entire way of life.

Figure 1-1. The four organs of mass communication. To reach the broadest audience, you need to move the process out of the *head* (1) and into the *heart* (2) with sincerity, into the *gut* (3) with humor and intuition, and, ideally, if you're sexy enough, into the *lower organs* (4) with sex appeal. Photo in the public domain; accessed via Wikimedia Commons.

The *heart* is the home for the passionate ones. People driven by their hearts are very emotional, deeply connected with their feelings, prone to sentimentality, susceptible to melodrama, and crippled by love. Religion tends to pour out of the heart, and religious followers feel their beliefs in their hearts. Actors usually have a lot of heart. Sometimes annoyingly so. In an episode of *Iconoclasts* on Sundance Channel, you can see it when Renée Zellweger (heart-driven actor) and Christiane Amanpour (head-driven reporter) visit the World Trade Center memorial in New York City. Renée is overflowing with emotion, crying for the people who died, agonizing over the tortured fate of humanity, practically throwing herself to the pavement in empathetic agony, while Christiane offers up analytical, dry-eyed, rational commentary on how sad it is that humans do terrible things like this (which she's seen firsthand all around the world in her reporting). It's a perfect side-by-side comparison of head versus heart.

The *gut* is home to both humor and the deeper levels of instinct (having a gut feeling about something). We're getting a long way away from the head now, and, as a result, things are characterized by much less logic and rationality. Humor tends to come from the gut, producing "belly laughs," but also is extremely variable and often hard to understand. There's nothing worse than someone trying to explain why a joke is funny.

People driven by their gut are more impulsive, spontaneous, and, most important, prone to contradiction. Where the cerebral types say, "Think before you act," the gut-level types say, "Just do it!" When things reside in the gut, they haven't yet been processed analytically. For that reason, when people have a first gut instinct about something, they generally can't explain why they have the instinct, where it comes from, or how exactly it works. As a result, if you quiz them about it, you're going to find they are full of contradictions. You'll end up saying, "But wait, you just said X is the cause, and now you're saying Y is the cause." And they will respond with crossed eyes and a look that says, "I know! Can you believe I'm so confused?" And yet they are still totally certain they understand what's going on.

We heard a lot about the gut-versus-head divide during the 2004 presidential race between George W. Bush and John F. Kerry. Bush even proudly spoke of how he based much of his decision making at the gut level. He told author Bob Woodward, "I'm a gut player. I rely on my instincts." Not surprisingly, Bush's presidency was characterized by a great deal of contradiction.

At the bottom of our anatomical progression we have the naughty *sex organs.* As soon as you finished reading that sentence, you probably smiled for reasons you don't even begin to understand. All I have to say is "penis" and you're either physically smiling or internally smiling. Why is this? Well, let's ask Bill Clinton—remember him? He's the man who obliterated his entire historical legacy thanks to this region. Let's ask the countless men and women who, over the ages, have risked and destroyed everything in their lives out of sexual passion.

There is no logic to the sex organs. Look at those arrows in the gut in figure 1-2. Now picture them moved lower and spinning in circles. You're a million

Figure 1-2. Intuition resides in the gut and tends to be full of contradiction. When the process is moved up to the head (intellectualized), the information is channelized, making it more consistent and logical.

miles away from logic in this region. And yet the power is enormous, and the dynamic is universal.

Not universal, you think? Some people have no sex drive? That is, of course, impossible to test, but one thing worth taking a look at is the life of the novelist and philosopher Ayn Rand. She was one of the most prominent popular figures to suggest it is possible not to be driven by such irrational forces. She authored the massively best-selling *Atlas Shrugged* in the 1950s and founded her "objectivist" school of thought and way of life on the principle of suppressing one's irrational side. And guess how her life turned out. She eventually got eaten alive by her sex organs.

Seriously. One of the greatest books I've ever read was Barbara Branden's biography of her, *The Passion of Ayn Rand*. In a nutshell, Barbara and her husband, Nathaniel, became followers of Rand, went to work for her, and believed and lived every word of her teaching about living an objectivist life—not allowing oneself to be controlled by pointless, frivolous, irrational thoughts and feelings. Rand's objectivist school of thought in the 1950s grew to enormous popularity; its followers even included former Federal Reserve chairman Alan Greenspan. And then . . .

Rand ended up secretly boinking Nathaniel for a couple of decades. When he dumped her, Rand turned vitriolic, and the public began to catch glimpses of the insanity she was living (proof that the story wasn't just Branden's fantasy). Total hypocrisy of the highest magnitude—telling the world to suppress its irrational side while viciously shoving the man who had scorned her out of her institute. According to Branden, Rand went to her grave still simmering with rage over it.

So don't even begin to think that the lower organs are not a universal driving force, for everyone from the local FedEx delivery guy to the president of MIT. And once you've processed that thought, you can appreciate the age-old adage "Sex sells." It's the truth, mate. If you are fortunate enough to get your communication down into that region, you can connect with almost every living human—even the most anti-intellectual NASCAR fan. Who doesn't like Brad Pitt and Angelina Jolie? They're sex-eeeee.

Too Heady: The Less Than One Campaign

Now, if we consider these organs, we start to see some fundamental differences in the members of the mass audience. The lower organs include everyone, but as we move upward, our audience narrows. There are people who pretty much respond only to sex and violence. Not much of a sense of humor, not much passion, and zero intellect. Once you move above the belt, you've lost them.

You still have the attention of a lot of people through humor—most folks love humor. But then you move higher and lose that element. Well, with the heart you still have actors and the religious folks. But then you move up above that, into the head, and who do you have left? Just the academics. Which is okay, but the point is that you're communicating now with a very small audience. You've left most of the general public out of the story.

So this is the fundamental dynamic. And it began to resonate with me in 2001 as I drifted back from the Hollywood environment I had been immersed in since leaving academia in 1994. I started working with academics and science communicators in ocean conservation. And as I did, the words of that acting teacher began echoing back at me.

I learned of a large project called the Less Than One campaign. The idea was built around someone's revelation that less than 1 percent of America's coastal waters are protected by conservation laws. Someone thought, "If we can communicate this factoid to the general public, when people hear it they will think about how small 1 percent is and they'll be outraged."

Well. They should have called it the Less Than Outraged campaign, since that's what happened with the general public. The Less Than One campaign opened its website in July 2003. It had a number of ill-conceived media projects (I'll talk about one of them in chapter 4), and, to make its short story short, by July 2004 the site was gone and not a trace of the project could be found on the Internet.

Suffice it to say, the masses simply do not connect with "a piece of data" (i.e., a number). Could you imagine a presidential candidate making his campaign slogan "More than 60 percent!" with the explanation that, if you elect him, eventually more than 60 percent of the public will earn more than $30,000 a year? For some reason I just can't see the crowd at campaign headquarters shouting, "More than 60 percent! More than 60 percent!" Sounds like something from a Kurt Vonnegut novel.

No, in fact groups connect with simple things from the heart—"A new tomorrow," "We've only just begun," "Yes we can." You just don't see a lot of facts and figures in mass slogans, unless they've been crafted by eggheads.

By now you may be thinking, "What's this guy got against intellectuals? He's calling them brainiacs and eggheads." Well, I spent six wonderful years at Harvard University completing my doctorate, and I'll take the intellectuals any day. But still, it would be nice if they could just take a little bit of the edge off their more extreme characteristics. It's like asking football players not to wear their cleats in the house. You're not asking them not to be football players, only to use their specific skills in the right places.

Kicking Flowers: The Value of Not Thinking Things Through

I'm criticizing overly cerebral people here, yet we obviously know there is a value to working from the head most of the time. Educated people make great inventions, create important laws, run powerful financial institutions. Clearly it pays to think things through so that everything is logical, fair, and consistent. But what's not so obvious is the value of sometimes *not* thinking things through.

Spontaneity and intuition reside down in those lower organs. They are the opposite end of the spectrum from cerebral actions. And while they bring with them a high degree of risk (from not being well thought through, obviously), they also offer the potential for something else, something magical, something that is often too elusive even to capture in words. And because they are so potentially effective, they are the focus of the rest of this chapter.

I learned about the power of spontaneity the hard way—by getting yelled at in that acting class. I eventually got to see it up close and personal as I began to realize I was a lousy actor. And the reason for my being a lousy actor was that I was . . . too cerebral. I thought too much.

Let me tell you specifically how I would get to see it. Night after night we would do acting exercises in which one person pretends to be at home and the other person comes home. On the edge of the stage was a fake wall with a door that the person coming home would enter. So, for example, I would be the guy at home, maybe working on balancing my checkbook, and my "wife" would come in after a long day of work. We would get into an argument over something, and then, right in the middle of the scene, I would accidentally do something that wasn't in the plan—like, let's say, knock over the vase of flowers on the table. The contents would spill all over the floor. I would look down. And then, being the highly cerebral former academic, I would start thinking.

I would think, "Wow, I just knocked over the flowers, that wasn't supposed to happen, we're supposed to be arguing over the wrecked car, how would this clumsy act I just did fit into my character's tendency to"—*and then, blaaaaah*, the teacher lady is up and screaming in my face: "*Stop thinking! Do something!* Nobody wants to watch you stand up here and think. You're like a statue. Do you want to watch a play full of statues? *Act!*"

Then a similar thing would happen with one of the younger, less cerebral guys. When he knocked over the vase, he would immediately kick it like a football and shout, "*I hate flowers!*" And the audience would burst out laughing and cheering, and the crazy acting teacher would scream at him, "*Why did you do that?*" and he would reply, "*I don't know!*" and she would scream with joy, because *that* was a spontaneous moment in which you could feel the magic.

And *that's* what I was so bad with. I would just think too much. The fact is, had she let me go long enough, I would have eventually looked at the vase and said to my "wife," "Your bad driving upsets me so much I end up doing things like knocking over vases of flowers." And the audience would have snored. I would have provided a well-thought-out and reasonable response to the spilled flowers; it just would have lacked that spark of energy that the other, more spontaneous performance provided.

That's the deal with spontaneity. It gives a wonderful energy that audiences love. And, by the way, it has become the core and backbone of a major shift in the entertainment world over the past decade.

The Shift to Unscripted Entertainment

I finished that acting class in 1996. I never had any intention of becoming an actor (I did it to improve my directing skills), but all the other kids in class headed off to pursue acting careers.

By early 1999, though, they began showing up on my doorstep, depressed. In Hollywood, the month of February is generally known as "pilot season." That's when the networks cast the pilots they will shoot—whether half-hour sitcoms or hour-long dramas. For actors it's a frantic time in which they may have four or five auditions a day, causing them to drive wildly back and forth between Hollywood and Burbank. But suddenly in 1999 the number of auditions dropped significantly, and my aspiring actor friends felt the pinch.

They would come to my apartment in Beachwood Canyon, right beneath the Hollywood sign, for lunch. We would sit on my front porch, and I would commiserate with them. "There are hardly any parts this year," they would say.

So where do you think all these acting roles went? Were they lost to outsourcing? Shipped overseas? Displaced by computer-generated actors? Nope.

They were lost to a new trend—reality shows, which are part of a larger category known as "unscripted entertainment." A whole wave of these shows hit the scene around the turn of the century, including *Survivor*, *Big Brother*, and all the other crazy shows you now know. But as quickly as my friends got depressed, they also heard a rumor that brought some relief—that it was only a fad—that within a couple of years reality shows would run their course, lose popularity, and never be heard of again.

Well . . . it's a decade later, and guess what? That rumor was way off the mark. Reality shows are as strong as ever, while sitcoms are officially a dying trend. Reality shows sounded the death knell for the sitcom; then another force, YouTube, came along and drove the spike deeper. Michael Hirschorn encapsulated this in an article in the *Atlantic* in November 2006 titled "Thank You, YouTube: DIY Video Is Making Merely Professional Television Seem Stodgy, Slow, and Hopelessly Last Century."

What do reality television and YouTube have that scripted sitcoms don't? Very simple—spontaneity. Or at least the feeling of spontaneity. Even though most reality shows do in fact have a very tight narrative structure, there is still something at the small scale, from one moment to the next, that feels uncontrolled, as if it has the potential to go anywhere.

Sitcoms, on the other hand, are controlled down to the very last detail. If a vase filled with flowers falls over, it's almost certainly because it was written into the script. Each show is broken into clearly delineated acts, with story arcs that follow standard patterns. The net result is an extremely predictable and formulaic style of storytelling. Having a strong, clear structure provides a level of comfort (we always knew Sam and Diane on *Cheers* would resolve their fight by the end of the episode), but eventually the predictability also leads to a loss of energy. The audience slowly absorbs all the major plotlines and standard setup/ punch line jokes until the whole genre loses its impact.

Spontaneity is fun, plain and simple. Just take a look at the annual Academy Awards ceremony—the Oscars. What does the public most crave every year? It's not the opening monologue, the dreary montages, the lame jokes from presenters, or the tedious musical numbers. What the audience desperately and eagerly prays for is the *one* spontaneous moment that will live forever. Whether it's Jack Palance dropping to the floor to do one-handed push-ups, Roberto Benigni hopping up on his chair as he calls to the stage, or Sally Field's "You like me, you really like me!"—that's what everyone lives for. It's the spark of magic that comes with spontaneity.

It's the same thing you can routinely see and hear at the Democratic and Republican National Conventions. The television commentators complain, over and over again, about the tightly scripted and controlled nature of the events. Every single moment, every speech, every presentation seems to be so tightly choreographed, down to the last detail. After a while, you get the feeling that the commentators are just hoping that someone, anyone, will trip on their way to the podium, interjecting at least one unpredictable, spontaneous moment.

If you want to see the truly blindingly brilliant charisma of a spontaneous moment, you should watch the original black-and-white film of President John F. Kennedy pinning a medal on astronaut Alan Shepard in the Rose Garden of the White House in 1961. Kennedy accidentally drops the medal, picks it up off the ground, and without missing a beat says, "I give you this medal that comes from the ground up," and the assembled crowd explodes with laughter. The scene has the sort of energy that political convention watchers dream of.

So what is it about spontaneity that is so powerful? It's the element of danger, the idea of performing without a net. These dynamics reach down into the lower organs—down to the gut with a twinge of fear.

And that brings excitement.

It also brings an organic element that has a feeling of truthfulness to it. That was the ultimate goal of the acting class I took with the crazy acting teacher—making the performance seem real. What she taught was a form of training known as the Meisner technique, named after legendary acting teacher Sanford Meisner. The training focuses on one main thing—getting you to the point where you are completely believable to the audience in the role you are playing.

It's also what improv acting is about: trying to create those electric, totally authentic moments, even at the expense of a lot of rambling, unfocused, less precise moments. Here's how this relates to scientists.

Over the past decade the science community has begun to develop at least some awareness that scientists communicate poorly and need help. Two major efforts to address this are the Aldo Leopold Leadership Program and the book *A Scientist's Guide to Talking with the Media: Practical Advice from the Union of Concerned Scientists.* Both are important projects, but both have their limitations in that they focus primarily on the first half of communication—substance— but don't yet reach much into the second—style. To explain this further, let me begin at the introductory level.

The Basic Principles of Science Communication

Science, from the beginning of time, has always consisted of two parts. First is the obvious part, the *doing* of science: the collecting of data, the testing of hypotheses, the running of experiments—all the standard stuff.

But there is a second part that isn't so immediately obvious, and that is the *communicating* of science.

Over the ages, *all* scientists, from the highest Nobel laureate to the lowest laboratory technician, have *always* had to take part in both of these activities if they wanted to actually be scientists. Even the technician who sits in the corner of the lab writing down numbers from the DNA sequencer has to, at the end of the day, communicate the data to someone. Without performing both parts (which happens all the time), you have not performed science. You get people who do the science and then fail to communicate it, and you get people who don't do the science but go ahead and communicate (the latter are known as frauds).

There are countless famous stories of great scientists who did a great job of the first part—doing the research—but then totally fell down on the second part. For starters, there's Gregor Mendel, the father of genetics. He is the true icon of poor communication. In fact, someone should create a Gregor Mendel Award for the scientist doing the best research yet failing to communicate it effectively.

Mendel was a humble Austrian monk of the mid- to late nineteenth century. While Charles Darwin was basking in the glow of the celebrity he had gained by communicating directly to the public with his best-selling *Origin of Species*, Mendel was toiling away in the Austrian Alps discovering the very genetics that would have given Darwin the mechanism of inheritance he needed to make his theory of evolution complete. But Mendel lacked the sort of self-promotional streak that is essential for scientific success in the United States today. He was a shrinking violet when it came to presenting his foundational work and instead published it in obscure journals before leaving this earth with little fanfare. His most important paper was cited only a handful of times over the next thirty-five years.

It wasn't until several decades later that a number of major evolutionists rediscovered Mendel's experiments and said to themselves, "Holy smokes, this guy worked it all out long ago." The rediscovery of Mendel led to what is known as the "modern synthesis," in which Darwin's ideas on evolution were brought together with Mendel's knowledge of genetics to create a robust theory of how evolution works. Had Mendel been a bit more of a communicator, the modern synthesis might have happened a few decades earlier and science would have advanced more rapidly.

A similar thing happened with Alexander Fleming, who in 1929 discovered penicillin but published his findings in a paper that drew little attention. Instead of going out on the road and communicating his discovery effectively, he left it alone and nothing happened for more than a decade. When Ernst Chain finally discovered his work in 1940 and heard that Fleming was coming to visit, he commented, "Good God, I thought he was dead."

Had Fleming's work been widely disseminated in 1929, it could have led to the development and application of penicillin a decade earlier, saving countless lives. Such are the costs of failed communication.

Effective communication is an essential part of science for at least two reasons. First, if nobody hears about your work, you might as well have never done it. And second, especially in today's world, if you don't communicate your research effectively, there are many people around who will communicate it for you, and when they do, it will probably be skewed in order to support whatever agenda they have.

The Objective/Subjective Divide

But if communication is so important, why don't scientists put more effort into it?

In my experience, it's because of the objective/subjective divide in science. The doing of science is the objective part. It's what scientists are most comfortable with. A scientist can sit in his or her laboratory all day long, talking to the microscopes and centrifuges, and they will never talk back. I have heard scientist friends of mine over the years rave about how much they enjoy field and laboratory research for exactly this reason—it's all so rational, so logical, so objective, and . . . alas, so nonhuman—a chance to get out in the field, away from people. No politics, no bureaucracy, no administrative duties, just pure rationality.

Unfortunately for them, there is that other part to science called communication, which involves dealing with those often irrational and illogical creatures called humans. And while Mr. Spock of *Star Trek* found humans to be fascinating, most scientists really don't.

In fact, in 1999 I did a video titled *Talking Science: The Elusive Art of the Science Talk*, in which I interviewed a variety of University of Southern California faculty members in the sciences, communication, theater, and cinema. One physicist told me about the whole syndrome in no uncertain terms. He said he had always, all his life, had a hard time speaking to people. So, when he went to graduate school to get his doctorate in physics, it was his dream come true to be paid to lock himself in a laboratory and not talk to anyone day after day. But then they broke the bad news to him—he would eventually *have* to go to a scientific meeting, stand in front of an audience, and give a public talk about his research. He was furious the day he learned of this, and at first he refused to do it. But it wasn't an option—it was a requirement. Over the years he has reluctantly taken part in the communication of his science, but to this day he says it's the worst part of his career. And I can assure you he is not alone.

Why is science such an antisocial profession? Is it that the profession selects for these traits, or is it that it reinforces these traits? Probably a little of each.

I think my moment of truth on this topic came in my first year as a professor, when I attended a big scientific meeting in San Francisco, scored a

Figure 1-3. The dual nature of science. The objective/subjective divide for both science and the communication of science.

poolside hotel suite, and organized a party in my room for the second night of the meeting. I invited about fifty scientist friends from the meeting, but when party time rolled around, about five showed up. All the rest either had evening sessions they wanted to attend or were getting ready for their own talks. I sat in my room that evening, staring out at the pool.

Scientists are wonderful people, but as a group they tend to be a little awkward when they get together. Going to the annual American Geophysical Union meeting just isn't quite the same as attending the Sundance Film Festival.

How can scientists overcome this? My theory is that they need to reach down into the lower organs. I begin by exploring the phenomenon of spontaneity.

How to Find Spontaneity

Not very spontaneous? Feeling like you're that guy who stares at the knocked-over vase and tries to think of what to say? Feeling like Chris Farley interviewing Paul McCartney on *Saturday Night Live*, where he mostly just stares at him and can't think of anything to say other than "That was awesome"? There are ways to work on this problem, one of which is called improvisational acting, or improv.

During my years in Hollywood I had several encounters with improv acting. For starters, I took classes at a couple of the improv programs that are scattered across Hollywood. In particular, I went through several levels of training at Second City, the program that gave rise to John Belushi, Dan Aykroyd, Gilda Radner, and many other great comics.

But more important, early on I became a fan of the legendary Groundlings School, located on Melrose Avenue in Hollywood. The school is one of the other prime training programs for the major comic actors who emerge on *Saturday Night Live*. It has its own suite of superstar alumni, including Will Ferrell, Chris Kattan, Phil Hartman, Paul Reubens, Jon Lovitz, Kevin Nealon, Maya Rudolph, Kristen Wiig, and many more.

After attending The Groundlings' Friday night shows for years, I finally broke the ice in 2002 by contacting one of the veteran performers, Jeremy Rowley, to see if he might be interested in helping out with my Shifting Baselines Ocean Media Project. I wanted to make a comic television commercial that talked about lowered standards for ocean quality by drawing comparisons to the idea of lowered standards for the arts. For one of the examples I wanted to have a scene of bad dancing. Jeremy had performed an incredibly funny scene in the Friday night show in which he ended up coming out totally naked holding a birthday party hat over his private parts and dancing to a frantic song from the Gipsy Kings. The scene produced screaming laughter from the audience—truly one of the funniest performances I've ever witnessed.

Jeremy helped me with the bad dancing scene, and then we put together a stand-up comedy contest for Shifting Baselines. We then cowrote and directed *Rotten Jellyfish Awards*, featuring Jennifer Coolidge (Stifler's mom in *American Pie*) and Daniele Gaither (of *MADtv*), followed by a series of comic short films using the main cast of The Groundlings. After that, I shot my *Tiny Fish Public Service Announcement*, starring Tim Brennen of The Groundlings and Cedric Yarbrough of Comedy Central's *Reno 911!*, and used a number of Groundlings actors in my feature films. So, over the course of seven years, I spent a considerable amount of time around The Groundlings and absorbed what I could of improv technique.

The most important overall aspect of improv training is that it is based on the idea of affirmation and positivity. (I talk about this in chapter 4, where I discuss the negating aspects of scientists.) But it also draws on spontaneity and the hugely likeable qualities that come with it. The object of improv is to work not from the head but from the gut. To listen very closely and to not wait for your brain to process what you're hearing, but instead to be guided by your instincts. Basically, to *trust* your instincts. To have enough faith in yourself that you don't feel the need to slow things down and think them through, but rather to simply act—impulsively, immediately, spontaneously. It's back to that kid kicking the vase.

Improv actors are like explorers—they open up doors and go inside. They do an improv scene in which someone comes out with something silly and nonsensical, and, instead of the other actors frowning and "negating" it by saying something like "That could never happen," they boldly move forward into uncharted waters.

For example, let's say the actors are pretending to be looking at a llama. One of them says, "Wow, look, it has seven legs." Instead of negating it by saying, "What? A llama could never have seven legs," another actor takes things in a positive, affirming direction by saying something like "Yeah, I wonder what happened to his eighth." And maybe the next one says, "Yes, llamas always conform to the rule of fours—this one must be a rebel." And onward toward increasing silliness, without a doubt, but also occasionally someone might nail a piece of logic. If there had been, for instance, a recent news story about a fast-food establishment having contaminated meat, one of the actors might say, in reference to the missing llama limb, "So that's what was in that fast-food meat." It doesn't all have to be baseless silliness, but it does all have to be affirmative because that helps the idea and the story get larger, and inevitably funnier.

In contrast, the scientist hears the "seven legs" statement and immediately says, "No, that's not possible," and the whole fun exercise crashes to a halt. Yes, this enters into the realm of accuracy, which is part of the scientist's job, but we'll get into that later, in chapter 3. For now, just know that the spark of spontaneity comes from not being careful, and it can be hugely powerful, as I got to see in my work with students.

No Joke: Improv Comedy for Scientists

In the same way that science splits into two parts—the objective (doing it) and the subjective (communicating it)—the communication of science has a divide. Looking back at figure 1-3, you can see there is the objective part of communication (the *substance* of what is communicated) and the subjective part of communication (the *style*). Knowing that scientists are drawn to the objective side of science, I think we can easily predict that they are also drawn to the objective side of communicating. And this tends to be much of the focus in workshops that train scientists to communicate better.

The Union of Concerned Scientists' book *A Scientist's Guide to Talking with the Media* asks, in the title of the fourth chapter, "Do you hear what you're saying?" It doesn't ask, "Do you hear *how* you're saying it?" It sticks with the *what*.

That's the difference: what = substance; how = style. Most teachers of science communication are still at square one, working primarily on the substance. And the idea of asking scientists to take lessons in comedy sounds rather absurd. But we've been experimenting with it at Scripps Institution of Oceanography with the graduate students and learning some fascinating things.

Every summer for the past few years I have taught the second half of the communication week in Scripps' orientation course for new graduate students. For the first two days of the week, the course brings in major print journalists from the *New York Times* and the *Los Angeles Times* to talk about communicating science from their perspective. They tell how to do a good job when you are being interviewed about your research or science-related issues.

In the second half we focus on electronic media, including an intensive video-making workshop where the students make their own sixty-second video. But a couple of years ago I decided to do a little experiment.

Some of the instructors at The Groundlings School, including Jeremy Rowley, occasionally run corporate training workshops in which they teach improv exercises to CEOs. They get them to work on lightening up and looking at their communication dynamics from a different perspective. So I managed to talk Jeremy into coming down to Scripps for a morning to do the same exercises with the students.

He ended up running two hours of improv games, which started out mostly silly, fun, and of questionable purpose—things like standing in a circle and taking turns saying the letters of the alphabet by having the person to your right look deep in your eyes and say his letter—"J"—then you turn to the person on your left, look deep in her eyes, and say, "K," and so on. Really just an icebreaker game.

But, as time went on, the games began to get more complex, and Jeremy ended each game with a detailed explanation of how it related to the students and their highly cerebral world.

The best game of all, and the one that brought the whole purpose home, was called the "add-on story game." Five students stood before the class. Jeremy chose one randomly. She began by making up a story—"Today my car broke down, so I had to take it to the shop." He interrupted her and randomly pointed to another student, who had to pick up where she left off. The next student said, "The mechanic looked under the hood, opened the carburetor, and found a dead bird in it." And then another student was chosen to pick up from there and keep the story going.

And *this* was where we got to see the true mind of the scientist at work. Some of the students kept their minds open, listened closely, followed the story. When called on, they instantly took their best shot at making up something that connected with what was said and kept the story going, even if their contribution sounded silly, like, "The bird woke up and flew out of the shop!"

But others—the more cerebral ones, the thinkers . . . ah, they were the ones who from the very start of the exercise went to work, thinking, "This is a story about a bird in a car motor. I'm eventually going to be called on at random. I don't want to embarrass myself, so I'd better have something prepared for when I get called on." Preparation, preparation, preparation—thinking, thinking, thinking. When they were finally called on, they would say something like "The bird had its wing stuck in the carburetor and couldn't get loose," even though the previous student (to whom they failed to listen) had just said the bird flew away.

And all of a sudden the story would stop dead.

The net result was very clear as the smiles vanished from everyone's faces and some of the students would say, "Oh, boo! No, that doesn't make any sense."

Jeremy would then stop the exercise and explain what had just happened. He would point out that the purpose of improv is, first, to listen very closely and, second, to trust yourself—to know that even if your mind is blank at the moment, you'll figure out something, even if it's as pointless as kicking the vase as the young student had done. And, finally, to do all you can to make your partner—the person who came before you—look as good as possible. Suddenly taking the story back to having the bird stuck under the hood makes the previous person look bad, as if he had been wrong in telling about the bird flying away.

You can see how this relates to being interviewed. In the one form of science communication training, you are told to arm yourself with a stack of sound bites, metaphors, analogies, and message points. Then, regardless of what the interviewer is asking, you are to push your own agenda and get *your* message out.

This orientation leaves the scientist thinking, "Me, me, me—I need to make myself look good." Which seems logical. But consider this—what if there is actually something unique to be gained by taking the opposite approach—by thinking, "Him, him, him—I need to make the interviewer look good"? Yes, it's counterintuitive. And so are a lot of things when it comes to communication, since it's not always entirely rational. Sometimes you need to be a little less direct and literal-minded (the subject of the next chapter).

With the improv approach, you try to make the interviewer look good. There is an upside and a potential downside. The upside is that you will have better chemistry in the interview, be more relaxed, a more enjoyable person. The downside is that you might not manage to "get in" everything you wanted to say or make certain everything is completely accurate.

Which is better? It's a question of substance or style. The former is better if you're in a setting where everyone is likely to hear and care about everything you have to say. But, if you're in a highly superficial medium like television, which is meant not for the academic audience but for the general public, and where people pick up much more on what they're seeing than on what they're hearing . . . then it's quite possible the improv approach will be more effective. It can result in the viewer saying, "I really liked that person who talked about global warming—she seemed really comfortable, knowledgeable, and . . . I didn't understand what she had to say, but just the fact that she seemed worried about global warming makes me think it's a serious issue."

That's in contrast to the scientist who spends the entire interview correcting the interviewer (i.e., negating), forcing the issues by giving answers that have nothing to do with the questions asked, and who seems to be pushing a story that the interviewer isn't asking for—something that happens every day on news shows.

For improv in general, the basic idea is saying, "Yes, and . . ." to everything that comes up.

Your partner says, "Look, there's Sasquatch, out in our front yard." You answer, "Yes, and . . . he looks really angry." Your partner says, "Yes, and . . . he just tossed your car over the house." And you say, "Yes, and . . ."

You just keep adding to the story, making it bigger and more interesting. You never halt the flow with anything negating—like "Sasquatch could never pick up a car."

It's a different way to communicate. It's not as precise as a scientist would like. But it is more likeable.

More from the Gut: Intuition

And now it's back to the battle-ax acting teacher. It's time for another one of her basic principles. This one is very powerful and leads us to the thing known as intuition. The concept is "Great actors memorize the script, then forget it." (Always made me think of those old denture ads, "Fixodent and forget it!")

That principle was repeated night after night, and it became very important to me years later. What it means is that, in the early stages, the actor ends up very much "in" his or her head, having just freshly memorized the lines. But with repeated rehearsal, the material gets committed at a deeper and deeper level—as if it drifts downward from the brain and into the lower organs. And as it does, the actor is able to add sincerity to the material as it moves down to the heart, then have fun with it and add more humor as it gets into the gut, and finally add genuine sex appeal when it reaches the lower organs.

But something extra happens when the actor "forgets" the script. After weeks of rehearsal, the actor goes away for a few days and doesn't think about the material. Upon return, the performance is no longer coming from the head. The actor is no longer standing in the room trying to picture the lines on the pages of the script. Instead, he is standing in the room, looking at the man pointing the gun. When he speaks, it comes not from memory but from what is seen and felt at the moment. It is alive and real. And—guess what—when he says, "Don't shoot me! I've got three kids," without even thinking about it, his words turn out to be very close, if not identical, to what the script said. When he "reaches" for the line, what he gets is what was in the script—available to him because the script was absorbed down at the level of intuition.

On a similar note, years ago I saw an interview with a British actor who was asked why his countrymen perform Shakespeare so much better than Americans. He said it's because British actors go beyond intellectual respect for Shakespeare. They are raised with the Bard from a very early age. By the time they are adults, they have committed the material to such a deep level that they are able to add all the elements of the lower organs to it—passion, humor, and even sex appeal. In contrast, American actors tend to learn Shakespeare later in life, treat it with overwhelming reverence and dignity, and end up "caught up in their heads"— still thinking, "Oh, my goodness, I'm doing Shakespeare; I'd better do it right."

Reaching into the lower organs is the ultimate goal of the Meisner technique, and it's what produces the wonderful, incredibly likeable chemistry that is the essence of good acting. This is what overly cerebral scientists lack—but it's an important part of interacting with the public. And it was a rule I tried to follow in making *Flock of Dodos*.

Dodo Intuition

In the spring of 2005, after running the Shifting Baselines Ocean Media Project for three years, I read about the conflict over the teaching of evolution versus intelligent design in Kansas and immediately decided I wanted to make a documentary about it. More important, I also decided to put to work all I had learned in my Hollywood education. Instead of studying the subject for the next six months, figuring out exactly what I wanted to say, and then writing a script, I wanted to rely on my instincts and get to work quickly.

Within two weeks of reading H. Allen Orr's article "Devolution: Why Intelligent Design Isn't" in the *New Yorker*, I was in Kansas with a film crew conducting the week of interviews that provided the core of the movie. Instead of carefully preparing for each interview, I opted to trust my instincts, trust my twenty years of studying evolutionary biology, trust my knowledge of editing (for ensuring accuracy down the line), and focus on doing a good job as an actor in each interview. I felt as if I had memorized "the script" over the past two decades. The best thing I could do now would be to forget it.

The result was that I didn't cover all sorts of important topics and questions that I probably should have in each interview. But the trade-off was that I was doing my best to listen to the person and respond, with as little thinking as possible, in an effort to generate good conversation.

This is an element of style that's difficult to teach in workshops and can be elusive to scientists who feel they owe their first allegiance to accuracy and the facts.

But there's more to life than just accuracy. Yes, that's a very touchy subject for scientists. Some might even disagree with that statement—saying that accuracy is *all* that's important. Suffice it to say, the topic is a major can of worms, which I will delve into in considerable detail in chapter 3. (Stay tuned!)

But for now, before moving on to the chapter's final topic—not being so cerebral—let me go back to that improv acting exercise at Scripps. It was such a fascinating contrast with everything the print journalists had taught in the first half of the week, and the students said so.

What the print journalists were teaching was substance—get all your facts organized, shrink them down to sound bites, figure out your message, go into any interview with a clear agenda of what you want to convey, and then make sure you are in charge. In fact, the Union of Concerned Scientists produced a PowerPoint presentation to go with its book on how scientists should deal with the media. It offers the following nine tips on preparing for an interview:

1. Do your homework. Before every interview, ask the reporter what the topic of the story is, where it will appear, and when and where the interview will take place.
2. Interview when you're ready. Even if the reporter is on a deadline, ask if you can talk in ten minutes so you can prepare your main messages and sound bites.

3. Repeat, repeat, repeat. Unless you are on live radio or television, every interview is edited. Take control of how you are edited by driving home your main points.
4. If you stray off course, bridge back to your main message.
5. End the interview on your terms.
6. Never speak off the record.
7. Never guess.
8. Emphasize qualifications (meaning if you have to make a point that has limitations to it).
9. Never get angry.

Let's take a look at these pointers and consider what sort of advice it is the authors are giving. If there's one basic principle they are espousing, it's that the scientist should control, control, control the interview. The first point says to *assert* yourself by insisting on knowing all the details. The second point says to *assert* yourself by not letting the interviewer start before you're ready. The third point says to *assert* yourself by making the same points over and over again. The fourth point says to *assert* yourself by bridging back to your main message. The fifth point says to . . . well, you get the idea.

It's nice that they're trying to instill self-confidence in scientists when dealing with the media, but take a look at it from the other side. If you were a journalist, would you want to be given a bunch of orders from the scientist you're trying to interview? "I'm not ready to start the interview. Let me make this point again. I want to say this again. Let me get back to my main message."

Finally, there is a danger to being overly prepared for an interview. A major television news reporter told me recently about an interview he did with a woman who is a top climate scientist. She showed up so heavily prepped, with her head so full of sound bites and analogies and catchphrases, that halfway into the interview she seemed to lock up—having a hard time connecting to his questions, giving answers that were so full of her message that they hardly related to what he was asking, causing him to have to ask questions a second time. She finally called the interview off, with much apology, saying it just didn't feel right.

The reporter told me he ended up so frustrated, wishing that she, and many other scientists he interviews, would just relax, trust him, and let him guide the interview instead of turning it into a struggle.

This is the divide between the heavy preparation and showing up with an agenda versus the improv style of trusting yourself. The former guarantees accuracy, and the latter leads to a much greater chance of hitting that one golden moment when interviewer and interviewee connect—the moment that later, in postproduction, causes the editor to turn around in his chair and say, "Hey, everybody, come take a look at this."

Take your pick which you'd rather have. Given that for television your one-hour interview will probably get cut down to thirty seconds, you begin to see the value of scoring that one great moment versus a solid hour of boring (but accurate) details.

Intuition

At the start of the chapter I mentioned the Myers-Briggs Type Indicator test. It is built around four "dichotomies," one of which is the divide between sensing and intuition. What this means is the split between people who want to base their decisions on information that is touchable, hearable, seeable, and present in the here and now, and others who are open to less tangible, more abstract information that could even be from the past or the future.

In essence, it's the same "head versus lower organs" divide I've been talking about. So, if the highly logical and analytical processes reside in the brain, what do we find at the other end of the spectrum?

Well, if we go way down to the far other end of the spectrum, we end up in the land of sex, and all hell breaks loose. This was Freud's undoing—trying to apply rationality to this realm. Good luck. He ended up with a career that was a mixed bag, which is why many scientists still despise him for coming up with nonscientific ideas—ideas that couldn't be tested or "falsified."

Basically, woe unto him or her who honestly thinks it possible to create rational and consistent theories of sexual forces. It's sort of like the observer effect, where you can never be certain whether what you're observing is the real state of nature or the state of nature that has been altered by your observing it. Same for sex. Those studying it have to deal with their own sex drives, which will probably drive them crazy.

Makes me crazy just to think about it. So let's stay away from this region. Use it at your own peril. Start off a speech with a sex joke at your own risk. Make a music video about the prodigious penis of the barnacle (barnacles have the longest penis relative to body size) and watch all sorts of weird things happen when you show it to groups of scientists (one male scientist accused me of being homophobic—how does that work?).

But there's another force, just above the belt, that is very important to science and scientists—intuition.

What is intuition? Start searching it on the Internet and you'll quickly find your way into wacky, far-out definitions like "the holistic merging of the cognitive senses," "the noncognitive experiences and memories," and "the body's bioelectrical sensitivities." Um, yeah. Right, dude.

Let's just say, in simpler terms, intuition is the act of knowing or sensing without the use of rational processes. Again, pretty much the opposite of what goes on in the brain.

Intuition is very important to the world of science because so much great science begins with it. There are countless famous examples. Descartes supposedly thought up the idea of Cartesian coordinates by lying on his back while sick, watching spiders spin their webs on the ceiling. Newton saw an apple fall from a tree. Kekulé dreamed of a snake biting its tail and came up with the circular molecular structure of the benzene ring.

These are all great discoveries that began as something that didn't look like science at all and lacked any data or rational thought. It's as if the gut is a great starting point for invention, innovation, or discovery. But once the idea begins to crystallize, it then must be transported northward to the brain so it can be subjected to the process of science.

James Watson described this interplay between intuition and science wonderfully in *The Double Helix*. In 1953 he and Francis Crick were getting close to discovering the structure of DNA and racing against a number of other scientists. Suddenly Linus Pauling at the California Institute of Technology beat

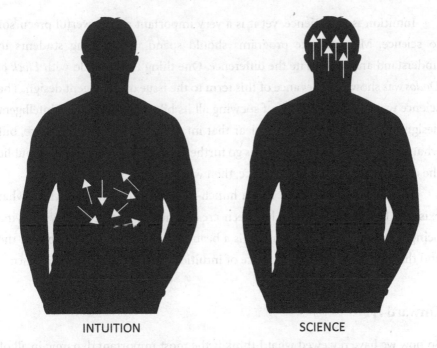

INTUITION   SCIENCE

Figure 1-4. *Left*, intuition: when Watson and Crick knew Pauling's structure was wrong. *Right*, science: when Watson and Crick had figured out how Pauling's structure was wrong.

them to the punch and published a paper with his version of the structure. They were stunned to hear the news, but Watson says that the moment he and Crick looked at Pauling's paper they knew he had it wrong. They couldn't tell you exactly how or why in that first instant, but their intuition made them feel certain they were right. It would take them several weeks in the laboratory to move their intuition up to their brains, formulate a solid explanation of why Pauling was wrong, and eventually come up with the correct structure of DNA, which is what won them the Nobel Prize.

For a much more detailed examination of intuition and its basic properties in the real world, read Malcolm Gladwell's book *Blink*. He talks about art forgery detectives who can spot a forgery almost immediately, but try to get them to explain why they know it's a forgery and you'll probably hear them offer up a lot of contradictory thoughts until they've had a chance to really analyze the artwork, move the process to their brain, and smooth out the logic and thoughts.

Intuition is not science, yet it is a very important and powerful precursor to science. More science programs should spend time getting students to understand and appreciate the difference. One thing I tried to do with *Flock of Dodos* was show the relevance of this term to the issue of intelligent design. The science world did a good job of spewing all its bile and anger at the intelligent design movement in making it clear that intelligent design is not science, but what few, if any, bothered to do was go further and answer for the general public the question "Well, if it's not science, then what is it?"

The answer is intuition. It is a hunch—a gut instinct that much of what exists in biological diversity has been created not by nature but by a divine being, the designer. This, to many, is a beautiful and inspiring idea, but in the end that's all it is—an idea—a piece of intuition. And intuition is not science.

Onward . . .

So now we have reviewed what I think is the most important dynamic in all of communication—the role of the four organs. And while we can plainly see that the brain is the epicenter for all that's permanent and lasting when it comes to information, I hope that you also now have an idea of what the lower organs can offer. They provide extra vitality, sparks of energy, an organic element—in general, they create the essence of what is meant by the word "human."

The tendency to be "too cerebral" leads to a preference to think rather than act (as in "doing something"). If you can manage to get past this and begin doing things, the next challenge is to find the creative energy to do the most effective things rather than just the most obvious. This requires that you not get too carried away with being literal-minded, as I will explore in the next chapter.

THEREFORE: UPDATE FOR THE SECOND EDITION

In this chapter I talked about the need to come down out of the head and draw on "the lower organs," which give rise to spontaneity and all things human. In the years since publication of the first edition, I've seen two clear patterns in the science world in relation to this. The first is the proliferation of improv training, which is exactly what's needed. The second is the National Academy of Sciences' effort to study and promote "the science of science communication," which is a disaster.

Making Improv "Mainstream" in Science

The folks at the State University of New York at Stony Brook did a very simple and effective thing for improv a few years ago by naming their science communication center after iconic actor Alan Alda. He has been a pioneer in addressing the need for improving science education, and as a veteran of Chicago's Second City improv school, he embodies all that improv is about. With the substance of the university's science communication center combined with the style of an improv actor, the topic of "science and improv" went overnight from sounding like a contradiction in terms to feeling like the cutting edge in better science communication. It was one of the best uses of the power of celebrity I've ever seen. Nowadays most science programs have at least looked into the idea of bringing in an improv instructor, which is great.

When I first started hiring improv instructors from The Groundlings School in Hollywood in 2005 (as I mentioned earlier in this chapter), the concept was still alien. It was fun to watch massively cerebral science graduate students get the chance to come out of their intellectual shell for a few hours. Back then none of them had heard the term "improv." Today they all have.

In the years since publication of the first edition, I've received a number of great e-mails from science folks who have heard what I had to say about the power of improv training. In fact, a year after the book came out I was contacted by Dr. Vicky Miller, a geneticist at the University of Bristol in the United Kingdom. She said that at their wedding, she and her husband read their vows aloud. He was a trained actor with a powerful voice, but she was a shy laboratory scientist with a quiet voice. She ended up disappointed that most of the guests couldn't even hear her.

Right about then she read this chapter, was intrigued by my descriptions of improv, and joined a local class. A year later she was finding herself, for the first time ever, feeling confident enough to stand up and ask forceful questions in her genetics department seminars. Improv can do wonders for introverts.

The Groundlings: The Gift That Keeps On Giving

For me, the best improv benefit of all has been working with The Groundlings. In 2004, I directed four short comic films for an ocean conservation event and got to know the talented main stage cast from the early 2000s. I cowrote the scripts with Jeremy Rowley (whom I talked about a lot in this chapter) and Roy Jenkins. Jeremy kept saying that some of the actors we cast would go on to become huge stars, as had Groundlings alumni Will Ferrell and Kristen Wiig. He was right.

We shot a crazy skit about a Senate hearing on coral reef decline that starred Melissa McCarthy; her husband, Ben Falcone; Jim Rash; and Nat Faxon. In 2011, the last two shared the Oscar for Best Screenplay for the movie *The Dependents*. That same year, Melissa McCarthy stole the show in the movie *Bridesmaids*, and the rest is box office history for her wildly successful career. You can find our silly little film on YouTube by searching for "Melissa McCarthy coral reef." Once you watch it, you'll never again be able to hear the word "day-tuh" and keep a straight face.

My collaboration with Brian Palermo has been the greatest gift from these collaborations with The Groundlings. Brian continues to be my go-to guy for improv. Let me tell you about three experiences in my work with him that leave me so deeply impressed with the training.

The first is that every time I do a workshop with Brian (which is at least a dozen times now), all the participants rattle off their names at the start, and I usually remember a total of about one of them. But he, more often than not, says, "Okay, let's see if I can do this," and then points to each individual and recalls his or her name. The first couple times I witnessed this, I thought he had a gimmick, like having studied them in advance. He didn't; he has just developed his ability to listen to an amazing level. Improv training does that.

The second example reinforces this. When I cowrote a book version of our workshop with Dorie Barton and Brian, we agreed to give each other notes on what we'd written. I went to breakfast with Brian, ready for a whole lot of delicacy (it's always hard giving notes to Hollywood folks), but I was stunned by his reaction. For every note, his answer was, "Okay, what if I try this—or what about this—or how about if I do this . . ." Basically, he listened closely and then gave the standard improv reply of "Yes, and . . ." It didn't matter how wrong or ill-conceived my note was. Never "Well, that's not a good idea." Always "Okay, sounds like you're thinking this, so how about if I try this . . ."

I know that everyone aspires to that sort of behavior, and I do my best to achieve it. But it was like having driven a really nice Mustang for years, being proud of it, and then having the chance to get behind the wheel of a Lamborghini. You don't know what is meant by "listening" until you have an experience like that. The Groundlings demonstrate how improv training enhances listening ability.

The third observation is the idea that "improv is like a muscle—you have to keep it in shape." Brian has kept his improv skills sharp for fifteen years by performing every Wednesday night with The Groundlings in "The Crazy Uncle Joe Show." A few years ago, the *Los Angeles Times* called the show "a pyrotechnic seven-person display of cleverness." If you're in Los Angeles on a Wednesday night, you should definitely catch the show. The cast is usually joined by a guest performer, like Will Ferrell, Eddie Izzard, or Jordan Peele. It was this "improv is like a muscle" principle that led me to the same realization for the narrative part of the brain.

Communicating Science Ineptly and without Intuition

All of this sounds as if science is moving toward becoming more human, since that's exactly what improv training does. Combine that with the March for Science and it feels like the needle might be moving. But right about the time it starts to look like the right changes are happening, I turn around and see the myopic mess that the National Academy of Sciences has created with a couple of symposia and a report that left me in stunned disbelief.

It started in 2011 with the Academy's symposium titled "The Science of Science Communication." When I saw the titles of the talks, all I could think of to say was "Why?" It's like running a symposium titled "The Violence of Gun Violence" in which you're advocating using violence to address the problem.

The biggest challenge to scientists, when it comes to communication, is to not be "overly analytical"—to not be too cerebral. You know, kinda like the title of this chapter?

Apparently the Academy so enjoyed all the talk about analyzing science communication in its first symposium that it felt the need to run a second installment. Then, as if to drive home the utter and complete tone deafness and blind-leading-the-blind dynamic, it came out with a "Research Agenda" report in the spring of 2017.

The report is not just misguided; it's a flat-out waste of money for whoever pays the bills. The report is titled "Communicating Science Effectively." Says who?

I searched the document for a number of words that are some of the most important in mass communication. For the words "intuition," "simplicity," "concision," "unify," and "understandable," I found zero mentions. Did you hear that? Zero.

But the psychology-oriented words "perception," "cognitive," "framing," and "complexity" were mentioned from nineteen to eighty-one times. There's your science of science communication—down the rabbit hole of psycho-babble, navel-gazing, and infinite complexity.

I called up the lead author, Dietram Scheufele at the University of Wisconsin. I wanted to know why there was so little focus on the intuitive side of communication. I wanted to know if the authors grasped the need to search relentlessly for simplicity—not just for simple-minded thinkers but because it is what all of science is about—the search for simple, singular unifying principles and theories and formulas and algorithms that reveal the structure in the universe.

All he really said, over and over again, was, "I think you and I are saying very similar things." To which I kept interrupting him to say, "No! Not by any stretch. What I am saying is not one bit similar to what you people are saying. You're sending out overly complicated messages into an overly complicated world full of overly stimulated students." What I should have said was, "You're not listening—you'll need to read the new chapter on listening in my book when the second edition comes out."

The fundamental divide is between intellectualism and intuition. The psychology stuff is intellectual and can be really cool . . . at cocktail parties. But communication in the real world requires intuition, which means simplicity and real-world experience.

Maybe I'm crazy. Maybe everything I'm teaching is wrong. The fact is, simplicity is everything when it comes to communicating effectively. Just ask Ernest Hemingway.

CHAPTER 2

Don't Be So Literal-Minded

Let's talk about the "science" of communication. Imagine that it's 2004 and you are at the Super Bowl and happen to be backstage just before the halftime show. You turn around and there's Janet Jackson, adjusting her black leather outfit. You say to her, "Hey, Janet, I'll give you a million dollars if you'll let me write my slogan on your breast." She recognizes you from high school (forgot to mention, you were best friends back then), giggles, says, "Sure, why not," and pulls back the flap over her right breast. You pull out a felt-tip pen, write "Save Coral Reefs" in big letters, and then, ten minutes later, when Justin Timberlake rips off her boob cover and all the cameras on the entire planet zoom in, you will have scored the greatest environmental communication coup in history. It's nice to talk about the "science" of messaging, sound bites, and focus groups, but . . . where would all that fit into what you just staged?

The Shortest Distance between Two Points

What is the shortest distance between two points? You know the answer. But what's the most effective pathway to take in going between those two points? Is

53

it always a straight line? If you're a scientist, you'll generally say yes to this. Why waste time and energy?

This is the way scientists tend to think. Because they are so analytical and caught up in their heads with the straightforward logic of the world, they tend to see things as fairly simple and direct. And if you want a good contrast, just think of posing the same question to an artist. The last thing in the world an artist wants to do is follow a straight line—how boring.

When I told my old science friends about the acting teacher screaming her head off at me that first night, their response was simple and literal-minded. They said, "She's insane. Just tell her she's incompetent and quit the class. You can find more level-headed instructors."

Well, yes. I could have. But . . . it just wasn't that simple. Obviously. I hung in there and went the distance, and here I am, a decade later, swearing by her teachings.

Here's an anecdote to illustrate this literal-minded phenomenon of scientists. When I got ready to hire an editor for *Flock of Dodos*, I gave my rough cut of the movie to a German friend, Pascal Leister, to consider. He called me up and said, "I only had to watch ten minutes to know exactly what this story is about. It's about the frustration scientists have in not being understood by the general public. I know this because my father is a nuclear energy engineer in Germany. I grew up listening to his frustration. He would walk around the house saying, 'Why don't they get it? We have all the data to show that nuclear power is completely safe nowadays, yet people just don't want to listen to it.'"

He went on to explain that the Green Party had done such an effective job of creating a fear-based communication campaign (reaching down into people's guts) about nuclear energy that scientists couldn't get the public to listen to their information showing that the actual risks are minimal. The result is that, while France is tripling its commitment to nuclear power, Germany is on course to get rid of it in a little over a decade.

Dreamland for Scientists

So the poor nuclear scientists are stuck with their literal-mindedness. They think that if you see a member of the public, all you have to do is explain the facts of nuclear power and assume that the person will listen and think. If only it were that simple—that would be a dream come true.

To illustrate the problem of literal-mindedness among scientists, I offer three examples, which I label "dreams"—as in "You're dreaming if you think it's this simple."

Dream One: Instant Messaging

I'm not talking about using your thumbs to send funny text messages back and forth on your cell phones. I'm talking about the delusion of literal-minded people that communicating to the mass audience is as simple as blurting out what you have to say—that you can put together your "message," say it, and people will instantly get it. Let me tell you about one of my run-ins with this. It's a story of people wanting to take an elegant message and make it disappointingly dull.

When marine ecologist Jeremy Jackson tracked me down in 2001, we began our film collaboration by making a short film on ocean conservation that was well received by the environmental crowd. Then, in the spring of 2002, he began telling me about a new term, "shifting baselines," coined in the mid-1990s by fisheries biologist Daniel Pauly. The term refers to the idea of losing track of the initial conditions of nature to the point where you can no longer accurately say how far nature has been degraded. Jeremy kept explaining it to me and saying, "Don't you think it's a very broad concept—almost the same as 'lowered standards'—that everyone who has lost track of the quality of their lives can connect with?"

I felt that was true, plus the term itself had a cool "ring" to it. And the clincher came when I explained it to my former film school friend Kevin Norton, who was working at a movie production company. He has no background in science, the environment, or any of the worlds I live in. I told him about it one evening at his apartment. He didn't seem to show any interest, and I just figured I had misfired.

But that night I got a call from him at 1:30 a.m. He was on his way home from a party, sounding happy, and said, "Hey, you know that term you told me about—I was just at this party and told these three hot chicks about it and they started saying, 'Wow, I *need* that term for my dating life—the guys I'm going out with these days are such schlubs, I've forgotten about how cool the guys I used to date were—I've shifted my baseline!' and they ended up telling everybody else and the whole night turned into a shifting baselines party!"

He called again the next day and said, "Hey, the guys here at the production company say you need to come over here and make a shifting baselines film about how things have changed here because nobody can remember the old days when they made really good shows—they've shifted their baseline!"

And thus the Shifting Baselines Ocean Media Project was born. The concept had been on tenuous ground prior to those two phone calls. It had been just Jeremy and me having a hunch that this could work as a communications theme. But Kevin's two phone calls removed the doubt for me. Not because it was an extensive scientific survey complete with focus groups and messaging analysis (like the Janet Jackson stunt), but rather because I could just hear it in his voice, and I sensed it as well. Sometimes that's better than a bunch of polling data. It's the old gut versus the head—which are you gonna trust?

So we scored our first $50,000 donation to begin building a mass communication campaign around this theme, we were on our way, and then . . . I had a rather unfortunate meeting with a "communications guru."

I'll leave this character nameless, but suffice it to say he had the environmental community at his feet, believing he was their visionary communications swami. He charged huge sums of money to conduct workshops where he pontificated on how to "message" and do other communications things that environmental groups yearn to do.

A friend knew him, thought he was great with mass communication, and figured the two of us would be a natural match. After my friend set us up to meet and told me more about this "guru" I, too, thought we would be a natural match. When I e-mailed Mr. Guru our newly developed Shifting Baselines materials, including my Sunday op-ed piece for the *Los Angeles Times*, which had already been accepted, I thought we would be a natural match. But five minutes into the meeting with him, I realized that, in fact, we were not anywhere close to a natural match.

He began with the term "shifting baselines," which he very, very confidently said was "too technical, too jargony, too sciencey." He kept saying, "Of course,

I know what it means, but most people wouldn't." He warned me that the final report of the Pew Oceans Commission would be coming out in a few months and would be sweeping the nation's media. He said that our little communication effort would only end up being "noise" to the general public that would just confuse them about what's going on in the oceans.

I kept trying to tell him it was already a done deal—we already had the first infusion of funding. So he ended the meeting by saying, "If you feel you have to do this thing, my one suggestion is that you change the name of the campaign to something more clear, like, maybe, 'The Oceans Are in Trouble' campaign. With that title, you'll never have anyone end up confused over what you're talking about."

Well, do you think that a year after creation of "The Oceans Are in Trouble" campaign, anyone would perceive it as different from all the hundreds of other "Save Our Seas" and "Protect the Oceans" campaigns? When a field gets crowded, there's a need to create a unique identity in order to stand out. That doesn't happen by making a plain, dull statement.

By then I had all sorts of friends around Hollywood talking about how much they liked even just the sound of "shifting baselines" as a name. To jump to the end of the story, seven years later the Shifting Baselines Ocean Media Project is still going strong, having produced a series of successful television ads, Flash videos, and short films.

But also, a little side note that really brings the syndrome to life. The following year I organized a stand-up comedy contest with my buddy Jeremy Rowley of The Groundlings. We got to thinking about how general this theme of "lowered standards" is for stand-up comics (we've all heard countless routines from comics talking about how far down their standards have dropped for dating, shopping, dressing, and countless other aspects of their lifestyles).

We ended up recruiting about fifty up-and-coming stand-up comics to deliver their best three minutes of material about lowered standards and shifting baselines. Wanting some ethnic diversity, I asked my friend Ifeanyi Njoku, a Nigerian filmmaker (and, years later, costar of my movie *Sizzle*), to recruit some African American comics as contestants. He called me a few days later and said, "It's not gonna happen—they say your environmental theme is 'too white'—they're all inner-city folk and can't relate to saving the oceans."

This was a disappointment, so I asked him to take a video camera and record what they had to say. He and his comic actor friend Alex Thomas (also a star of *Sizzle*) went down to Crenshaw Boulevard, in the heart of South Central Los Angeles, and walked up to randomly chosen African American folks to quiz them about this. Ifeanyi and Alex tossed out keywords and got all sorts of hilarious responses—especially for the word "Greenpeace"—for which people replied, with zero hesitation (not as if they were trying to craft a clever answer), "Marijuana," "The chronic," "That bomb-a** sh*t that knocks you on yo' a**."

When they asked people, "What comes to your mind when I say the term 'shifting baselines'?" the responses were clear and immediate—again, no confusion or hesitation. The first guy said, "That's like in rap music when you shift the bass line." The second guy said, "Oh, that's like in basketball when you drive the baseline." And the last guy, standing out in front of a fast-food joint, said, "Well, you see, the baseline is like the waistline—you gotta grab her by the hips and shift her baseline when you hit that a**," as he made the gestures of having sex.

The point of the entire exercise was that it showed how non-"sciencey" the term "shifting baselines" was. Had the name of our project been "Hyperstatic Metastable States," when these people on the street heard it they would have looked cross-eyed. But it wasn't. And not only did the "baselines" part resonate with sports culture, the word "shifting" has been the central theme of Nissan's car commercials (shift your horizon, shift your perspective, shift your driving) for nearly a decade. Both words are in the zeitgeist. Kevin and I somehow knew that—actually, it's more as if we intuited it. The communications guru didn't. And he, I'm afraid, just ends up being another example of the blind leading the blind in the communication of science and environmental issues.

Getting back to the original point of this chapter, there is a spectrum for any given piece of information, stretching from boringly blunt to incomprehensibly elusive. Naming a campaign "The Oceans Are in Trouble" is the former, as dull as dishwater—nothing intriguing, nothing inviting, nothing that arouses the curiosity—while "Hyperstatic Metastable States" is the latter—so technical that the public is lost, with nothing to grab onto. The object is to find something, like "shifting baselines," that falls right in the middle. Elusive enough to sound a little intriguing but familiar enough to roll off the tongue.

And this is where the overly literal-minded thinking of the scientist falls short for mass communication. The "Why waste time? Just tell 'em what it is" philosophy may work for science students, but it doesn't for the broader audience. This is where so much of the art of communication resides.

Others besides the communications guru have questioned whether this sort of indirect communication works with the public.

In 2005, Ken Auletta addressed this issue in a *New Yorker* article titled "The New Pitch: Do Ads Still Work?" He talked about the breakdown of traditional advertising markets with the advent of new media—how it's left old-timers adrift and many questioning whether advertising still works. But, after examining how complex it's all become, he brings it back down to simple elements with the story of an ad executive who, while walking to lunch one day, kept repeating his company's name until he finally realized it sounded like something a duck would say.

Thus was born the highly successful ad campaign for Aflac, built around an obnoxious duck quacking the company name. But, as Auletta points out, in the beginning there were literal-minded skeptics who thought the concept was a terrible idea for an advertising campaign because a duck quacking said absolutely nothing about the attributes of the insurance company. And yet . . . four years later, without making any major changes in its business practices, Aflac had doubled its sales.

Someone should have said to those skeptics, "Don't be so literal-minded."

Of course, it's difficult to know what's too elusive and what's too literal. When we were preparing to take *Flock of Dodos* to the Tribeca Film Festival, we interviewed a number of film sales representatives—the people who serve as agents in selling your movie to distributors. One fellow loved the movie, but there was one hitch—he didn't like the title and wanted to change it. His reason was that he didn't like titles that don't tell you what the movie is about. He wanted something more literal, like *Unintelligent Design*. But we had already created the opening animation of dancing dodos, which everyone loved, so there was no changing the title. It was a deal breaker. A few days later, I realized how passionate the guy was about this issue when I looked again at the name of his company—it's called the Film Sales Corporation. Guess what they do.

The title of a product often ends up being a direct struggle between the marketing voice and the artistic voice. I once had the wonderful experience of meeting a famous political activist from South Africa, a black man named Don Mattera, who had spent years in prison and engaged in the struggle against apartheid. He had published a best-selling book in South Africa titled *Memory Is the Weapon*. With such an evocative title, you can imagine how powerful it was. But he told me that when an American company picked it up for distribution, its marketing people changed the title to *Memoirs of an Apartheid Protester*. That's called stomping the life and beauty out of a piece of art. But if I had been one of the marketing people, I'd have probably recommended that title to make sure it sold at least a few copies in the United States, where no one knew of the author.

Dream Two: Instant Victory (by Going for the Jugular)

"It was okay, but we were hoping he would have 'gone for the jugular' more."

That was what friends told me a number of major evolutionists had to say about *Flock of Dodos* in 2006. They were expecting the movie to be an all-out assault that would use "the facts" to grab the intelligent design movement by the throat and use information to slit it, right at the jugular. As if that were actually possible.

In the end, it was another aspect of the hazards of literal-minded thinking. In fact, it was this sort of thinking that had drawn me into making the film in the first place.

In May 2005, as I got started on the movie, I began calling my old evolutionist friends and listening to their stories. One thing I heard from a number of them was that evolutionists were conducting public debates against proponents of intelligent design and consistently losing. And they were furious about this.

I spoke with one of my old buddies, a professor of evolutionary biology at a small college in Massachusetts, who told me he had debated an advocate of intelligent design and lost. He said he had stood on the stage and laid out the facts showing how there was no way intelligent design could possibly be tested and falsified and therefore was "not science" and shouldn't be taught in the classroom. He had been proud of himself as he finished his presentation and looked at the audience. But all he saw were expressions of resentment and looks that seemed to say, "You think you're such a smarty-pants." He was beginning to learn that sometimes it's not as simple as "see audience, hurl facts at audience."

His opponent had scored much better, not by arguing "the facts" but by issuing impassioned, heartfelt pleas about the need to teach students "critical thinking" and, supposedly, to support the American principles of freedom of expression by allowing students to question whether evolution and "Darwinism" were really the airtight bodies of knowledge they supposedly claim to be. Most important, he came across as more likeable by not being overly confident and so caught up in "the facts." His approach was less literal but ended up being more effective.

The evolutionists were going down in flames on the stages of America, and it got bad enough that the National Center for Science Education eventually established a simple recommendation, which was never to debate a creationist or an advocate of intelligent design in public. The main reason for this is the imbalance between evolution, which is largely intellectual in nature and thus comes from the head, and creationism and intelligent design, which, being religious in nature, come from the heart. Just recall the four organs concept and you can see the automatic difference in audience size—it's just not a fair fight. The heart audience is bigger. And in fact if you look at the size of the major evolutionary science organizations in this country (probably tens of thousands) versus the size of all the churches (millions of members) . . . well, case closed.

But there is also this dynamic of literal-minded scientists believing it's just as simple as slaying them with the facts. My Shifting Baselines partner Steven Miller, a marine biologist, is fond of saying that the science establishment often thinks it just needs to "argue louder"—meaning use the same fact-based approach, just more forcefully. And this was the situation with the major evolutionists: they wanted me to damn the intelligent design believers with data. But it's just not that simple.

There is no jugular to go for. If there were, someone would have managed to have severed it long ago. In the great "Scopes Monkey Trial," famed lawyer

Clarence Darrow used wit and logic to out-argue William Jennings Bryan. And, though the fictional portrayal of the great creationism/evolution debate presented in the play *Inherit the Wind* made it seem like the forces of "the head" won in that conflict, the fact is that the court ruled against Scopes and creationism continued to spread through the land like wildfire.

This is not to say that there aren't ways to defeat heart-driven creationists in a debate—there are, and I'll get to that later. But what's most important is to realize it takes more than just the straightforward blurting out of facts.

Dream Three: Instant Enlightenment

I like to refer to this tendency to believe that "the facts speak for themselves" as "science-think." It is truly wishful thinking, though every once in a while a single, simple idea is enough to catalyze change. If you find the cure for a disease or create an amazing invention, you don't really need a public relations campaign to make it take off, at least somewhat. And yet, in America, even the best ideas still need help. That help comes through communication, whether it's an advertisement for a product or news stories about a project.

But scientists fall victim to the belief that information alone is enough to effect change. They think, "If I can just put these facts together into this specific argument, when people see it all assembled they will change their outlook." Which might be true *if* people actually see it. But that's the problem.

For a specific case study in this problem, I turn to the final report of the Pew Oceans Commission, released June 2, 2003. It's a textbook case of a group of technical types who succumbed to science-think. They thought they would be speaking with a loud and powerful voice, but in the end, they didn't.

In the spring of 2003, as we began to get our Shifting Baselines Ocean Media Project off the ground, with plans to film our first television commercial using comic actors, I began to get pushback from the ocean conservation world. A number of communications directors for ocean conservation groups told me about the "firestorm of media attention" that was about to be unleashed. "The final report of the Pew Oceans Commission is coming out, and when it does, it's going to take the nation by storm," said one ocean communicator. "It will be the lead on every evening news channel, will be on the cover of *Time* and *Newsweek*, and will be the talk of the nation. The conclusions of the report are devastating."

The Pew Oceans Commission's study was a three-year project with a budget of $3 million. The goal was to pull together the first overall assessment of coastal waters and resources of the United States since the Stafford report of the early 1970s. By the time of its completion, the head of the commission, former White House chief of staff Leon Panetta, was saying that when the American public reads the report they will be "angry" about the devastation of our coastal resources.

Well . . . maybe they would have been. If a lot of people really had ended up reading it. But they didn't.

As the date of the report's release approached, I began to hear from insiders that they had made a little tactical error. No money had been set aside in the budget to create a media campaign around the findings of the report. I spoke on the phone with Justin Kenney—one of the only staff members left in the commission's offices as it ran out of money—who said, "I'm not sure we've even got enough money for coffee at the press conference."

A press conference was held, colorful brochures for the final report were handed out, and the members of the commission were present and suitably outraged over how bad the deterioration of our oceans had become. But when the news of the report hit the media, it ended up not on the cover of *Time* and *Newsweek*, not on the evening news, not even on the front pages of the major newspapers, but on page A22 of the *New York Times*. Instead of a big bang, there was hardly a whimper.

After the press conference, the members of the commission went home, returned to their normal lives, and, by all accounts I heard, only one of them, Roger Rufe, who at the time was president of the Ocean Conservancy, found time to go out on the road and give a few talks in which he tried to convey to the general public how important this study was.

What happened? Science-think. The heavily academic-cerebral-scientific makeup of the commission (most members had advanced degrees) led them to believe the information in the study would be so immediately compelling, so jaw-droppingly profound, that it would sell itself by word of mouth. They believed that journalists would sit down, read the entire thirty-five-page document, and feel their world had been shattered, causing them to stay up all night producing urgent and compelling media. Instead, what they got was a collective yawn.

Did the Pew Oceans Commission's report have to be such a dud? Is it just the case that any report is destined to be greeted with minimal interest? No. The final report of the 9/11 Commission the following year showed that such a study can have a significant impact *if* it's accompanied by a solid communications effort. The *New York Times* reported on July 19, 2004:

> Members of the independent Sept. 11 commission say they will mount an aggressive nationwide lobbying campaign to pressure the White House and Congress to overhaul the nation's intelligence agencies, an effort they say will begin this week with release of a unanimous final report criticizing virtually every element of the way the government collects and shares intelligence.

The lobbying effort would be a break with tradition, since blue-ribbon federal commissions often disband almost as soon as they have completed a final report, the members returning home from Washington and leaving the report to speak for itself.

Granted, more Americans may be interested in terrorism than in the oceans, but the 9/11 report didn't sell itself. It was released as both a hardcover book and a book on tape, and the committee members toured the country, lobbied, and eventually testified before Congress. Two years later they barely, just barely, managed to implement some of the report's recommendations—which just shows how hard it is in the United States to bring about change. But it is still possible if you couple the product with an effective communications effort.

And notice that last sentence—the part about the members scurrying home and leaving the report behind to speak for itself. Most committees do that and don't care. What was worse with the Pew Oceans Commission's report was that the folks associated with it actually deluded themselves into believing that their report's "speaking for itself" would be enough to make an impact.

A Turnable Tide?

This mode of science-think is deeply ingrained in the academic world in all different aspects. A 2003 study funded by the David and Lucile Packard Foundation revealed this for the world of ocean conservation. The team of environmental policy analyst David Wilmot and environmental lawyer Jack Sterne produced *Turning the Tide: Charting a Course to Improve the Effectiveness of Public Advocacy for the Oceans*, which dug to the bottom of this syndrome.

They examined why the ocean conservation movement has, in general, fared so poorly while others have succeeded. For comparison, they looked at relatively successful lobbying groups such as the National Rifle Association and Citizens for Tax Reform. And they came to a simple and blunt conclusion—that ocean conservationists, in general, tend to be drawn more to policy than to politics.

In other words, the powers that be in the ocean conservation world tend to be more comfortable with writing new laws, funding more studies, gathering more data, and sponsoring more workshops, in hope of putting the pieces together into that magical argument that will spring to life by itself, than with going to Washington, DC, and using money to hire lobbyists the way the big boys do it. They would rather stick to the objective elements (science, law, policy) than dabble in the subjective elements (communicating, lobbying, persuading).

I personally witnessed this trait in central California when I tried to get the ocean conservation "powers that be" to fund short videos explaining the overfishing issues to the fishing community and the general public. Representatives of a major funding organization eventually said to me, verbatim, "Why should we fund your communication efforts when we have the lawyers and the legal system on our side to simply override these fishermen?"

That's called the brute force approach to getting your way, which certainly can win, but it's a bit like leveling the village rather than trying to win hearts and minds through communication.

Suffice it to say, this aversion to communication, whether it's intentional or just the result of cluelessness, is real and can be quantified. The business community long ago figured out that you need equal efforts allocated to the objective part of your project (the creation of the product, whether it's a government report or an automobile for sale) and the subjective part of your project (communicating to the public that you have something they should check out, also known as advertising).

In Hollywood, this revelation hit home in 1977 with the release of the first *Star Wars* movie. I remember reading initial reviews saying that this was a weird and unusual movie that no one was certain would succeed, since there seemed to be a sort of cowboy western corniness to it. But *Star Wars* was the first true blockbuster movie to be accompanied by an enormously expensive marketing campaign. As money was spent on promotion and massive advertising of the movie, the skeptics sharpened their knives. But when the box office began to skyrocket, the entire industry was turned on its head, never again to be the same.

Decades later, by the summer of the Pew Oceans Commission report, the result of this shift in paradigm could be seen with any given movie. Figure 2-1 shows three representative movies from that summer. The proportional amounts spent on advertising range all the way up to *Napoleon Dynamite*, which was produced for only a few hundred thousand dollars, but the distribution company, knowing it had a winner on its hands, gambled $10 million in advertising—equaling 96 percent of the total budget for the movie.

I remember hearing quirky radio ads about the movie long before it arrived at theaters, playing sound bites of Napoleon saying, "Gosh, why do you have to be so stupid?" And the result of that gamble? Over $50 million at the box office, followed by a massive DVD release. The Hollywood folks know what they're doing in terms of following through with their products. They know that "the product speaks for itself" is not a sufficient business strategy.

In contrast, we have the Pew Oceans Commission report, which had for its final communication effort about 3 percent of the entire budget. Is it any wonder that it failed to have much effect? Need I say more about science-think? The world has changed. The public no longer sits in patient silence, awaiting word from the science community. It's a tough marketplace, but not impossible, as shown by both Hollywood movies and the 9/11 report.

This is the dilemma of science-think and yet again a situation in which scientists simply shouldn't be *such* scientists. Bring in the professionals, and trust

Figure 2-1. The relative expenditure on creation of a product versus the marketing of the product. The first three pairs of bars are representative Hollywood movies in the summer of 2003, for which the expenditure on marketing ranged from 38% to 96%. The last pair show the Pew Oceans Commissions report, released in 2003, for which about 3% of the budget went to marketing (i.e., communication). Needless to say, the Pew report had the societal impact of a Hollywood dud.

them when they tell you to invest in communication. It may be frustrating and seem like a frivolous waste of resources, but what's the alternative strategy—to assume that people are rational, thinking beings? There's a famous quote by Democratic presidential candidate Adlai Stevenson, who heard a woman shout to him that all the thinking people of America were with him. He replied, "That's not going to be enough, Madam; I need a majority of the public."

And if you have any real doubts about the extent to which the United States is a media-driven society, first read David Halberstam's brilliant *The Powers That Be*, a book that changed my life, and then read Jerome Groopman's 2006 *New Yorker* article "Being There." Groopman reports that just 15 percent of emergency room resuscitations in the United States succeed in saving the patient, but surveys of the general public reveal that, because of what they see on television, they expect about two-thirds of all resuscitations to be successful. Television shows like *ER* and *Rescue 911* tend to tell uplifting stories with happy endings, skewing our perception of the success rate. It goes to show, once again, that we live in the land of media, where the boob tube is a major source of (mis)information.

Solution: Remember the Octopus

The whole idea of nonlinear thinking can be extremely difficult. It runs so counter to the notion of just seeing and responding. It makes me think of an encounter with an octopus a marine biologist friend of mine had many years ago.

He and I were part of a four-man team who spent a week living at a depth of sixty feet in the Hydrolab Undersea Habitat, which was operated by the National Oceanic and Atmospheric Administration in St. Croix, U.S. Virgin Islands, until the late 1980s. On the last night of our stay in the habitat, my friend and another member of our group went out for a night dive. While out on a sand flat at a depth of about eighty feet, they encountered a substantial-sized octopus— about three feet long.

My friend wrote on his dive slate the famous last words, "Get a photo of me with this thing," and then reached for the octopus.

Cut to me sitting inside the habitat, talking on the radio to the shore base as I see a blood-covered hand break the surface of the water in the hatch that leads into the habitat. My friend surfaces, calmly saying, "It's only a flesh wound," and then reveals a one-inch gash on his wrist that has already swollen up to look as if a golf ball had been inserted under the skin (the localized response to the venom injected by the octopus).

The next day his dive partner reenacted for us what had happened. After my friend picked up the octopus and posed for the photo, the octopus began clamping down on his arm. The photographer said he saw the smile vanish from my friend's face and a look of terror become visible through his face mask. He had shifted from posing to panicking. When he tried to pull the beast off his arm, the octopus responded by simply clamping down harder. The more he pulled, the more fiercely the octopus held on. Until finally . . . the water was filled with a bloodcurdling scream from the diver as the angry octopus expressed his annoyance with his parrot-like beak.

The diver kept fighting, but the octopus remained clamped on. It wasn't until he finally relaxed and quit pulling on the octopus that the animal let go and swam quickly off into the dark (with an awesome story to tell his friends).

The moral of the story is that sometimes the reactive response just doesn't work. The diver fought and fought, trying to get free of the octopus, but the harder he fought, the tighter the animal held on. It wasn't until he finally did the counterintuitive thing—relaxed—that he at last got what he wanted.

And that's how it works with communication. Sometimes, particularly with the mass audience, people don't want their information told to them directly. You can pound them with the facts all you want. They're just going to clamp their hands over their ears until finally you figure out a more indirect pathway to their brains.

Just take a look at *The Daily Show with Jon Stewart*. In 2008 a critic for the *New York Times* posed the question "Is Jon Stewart the most trusted man in America?" And she wasn't joking. *The Daily Show* has gained enormous popularity, and audiences increasingly view it as a source of serious news. But the packaging of the information is incredibly indirect and surrounded by all sorts of nonsense and silliness. Yet in an information-saturated society, it ends up being a popular source of news for many.

Whether it's a television audience or an angry octopus, sometimes the answer isn't sheer force or straightforward facts. It's something less literal.

A Powerful Concept: Arouse and Fulfill

A pattern is now emerging in these basic dynamics. Most of what I've been talking about has two components—objective and subjective, direct and indirect, literal and nonliteral.

I've talked about how science has two parts, the doing and the communicating. I've talked about how communication has two parts, substance and style. And I've talked about how successful politics and business have two parts, the production of a report or product and the lobbying or advertising of it.

The two parts—the yin and the yang, the here and the there, the ebb and the flow. And once you look at things this way, you begin to see the dynamic everywhere. Which was the case for me when I made a video about giving scientific talks.

I emerged from the USC School of Cinematic Arts having finished my master of fine arts degree in film production. I had left the academic world behind in 1994; most of my old colleagues had written me off as having had a midlife crisis and never expected to see me again. I had immersed myself in the world of filmmaking, acting classes, Hollywood parties, movie premieres, and surfing. Academia had become a distant memory.

But in 1998 a group of my old science friends invited me to a symposium in Denver that was a tribute to Alan Kohn, one of my favorite professors from my undergraduate days at the University of Washington. They asked me to come and speak about my filmmaking in the middle of their science meeting. So I went. And I saw all these folks I hadn't seen in five years. But more important, I sat through an entire day of twenty short talks by scientists about their current research.

The talks were exactly the same presentations I used to give when I was a scientist—people standing at a tiny screen, backs to the audience, pointing to their cluttered, confusing slides (this was in the days before PowerPoint), muttering "Um" with every other sentence, and rambling on with no beginning, middle, or end to what they were saying until finally the moderator cut them off by saying, "You're out of time," at which point they would snap out of their droning, turn to the audience, and say, "I guess that's about enough for today." I sat in the back listening, staring in disbelief, and asking myself, "How in the world did I ever sit through these talks?" By then, my little academic brain had been subjected not just to the screaming acting teacher but also to dozens of classes in shooting and editing film. I had watched films so hyperkinetically supercharged that they could reach inside your eyeballs and make your visual cortex vibrate.

In quantitative terms, I had spent five years working in the world of thirty frames per second (which is what you get with video). If a frame is a picture, and a picture tells a thousand words, you could say I was living in a world of 30,000 words per second.

And suddenly here I was, sitting through slow, monotonic, often image-less presentations with a pace of maybe two words per second. It was like a baseball batter being thrown a change-up pitch.

I returned to the University of Southern California and told some of my friends in the biology department how bad these presentations really are. They replied, "We know. We're scientists; we're terrible at giving talks." I told them about some of the nifty things I had learned in film school about visual expression and the need for simple presentational elements such as sufficient screen size and sound quality for audiovisual presentations. They were intrigued, and so they pulled together some funds for me and I set to work on a twenty-minute video to explore this. The video was called *Talking Science: The Elusive Art of the Science Talk.*

I interviewed a number of faculty members from the cinema school and the theater department, a number of science faculty members, and—most important and interesting to me—several faculty members from the USC Annenberg School for Communication. A decade later, among all the interviews in that video, a single sound bite stands out, which I think gives one of the most powerful general rules for all of this communication stuff.

It came from Tom Hollihan, a communications professor, who said simply and elegantly, "When it comes to mass communication, it's as simple as two things: arouse and fulfill. You need to first arouse your audience and get them interested in what you have to say; then you need to fulfill their expectations."

And that's about it. Motivate, *then* educate.

When you begin to digest this, you realize that most failed communication efforts are the result of falling down on one side or the other. Academics tend to fail to motivate; they just jump right into the fulfillment part. I can't tell you how many times I've asked scientists what they study and been immediately bombarded with every little detail of their research.

Conversely, Hollywood makes the other mistake, getting an audience fired up and then failing to deliver any substance. The classic flaky Hollywood environmentalist is filled with passion and can get others equally upset about an issue. But then, when you want some specifics about exactly what's going on, you get a bunch of heartfelt nonsense. I suppose it was a little mean of me to parody this weakness in *Sizzle,* but it resulted in one of my favorite lines in the movie: the producers turn to me and say, "We feel very, very passionate about global warming, and we're very, very upset about it. We just don't know why we're upset."

Academia and the Prearoused Audience

Back when I was a professor, I was proud of my oratorical skills. Students would hang on my every word. Or so I thought—until, on an overnight field trip to Maine, I spent much of the drive up telling the students in the van all sorts of exciting stories of my days in Australia studying the Great Barrier Reef. They couldn't seem to get enough of my tall tales.

But that night, we were camped out in tents and I overheard a group of the students who didn't realize I was right outside the tent. They were laughing about how they had been encouraging me to tell more stories because they thought I was having fun doing it, and that would make me like them better and thus give them better grades. That was my first awakening to the fact that students are not a realistic audience.

In fact, to the contrary, they are a "prearoused" audience. They walk into classrooms and lecture halls already "interested" in the material, not because the lecturer is a brilliant speaker but simply because they want a good grade. I know this sounds cynical, and no doubt a lot of the students are genuinely interested, but seriously, let's be honest about this. Most of them are there to get a good grade. The professor doesn't really need to waste time and energy "arousing" them, and as a consequence many professors don't.

The result is that professors tend after a while to look out at that sea of attentive faces and think, "Damn, I'm a good speaker—these students eat up everything I say."

I can't tell you how many of them I've known over the years who have fallen victim to this. And it's a shame because occasionally you come across the ones who aren't fooled and who do realize you need both elements, and those professors are magic to listen to.

Within my limited experience, the most wonderful of them all was also, in my opinion, the greatest evolutionist of modern times, Stephen Jay Gould. In graduate school at Harvard University, I was fortunate enough to be a teaching assistant in the introductory biology course he taught. He was a fantastic lecturer. And you can still see his gift for both sides of the formula—to both arouse and fulfill—laid out bare in every single monthly column he wrote for *Natural History* magazine for more than twenty-five years.

Every one of his columns began with a few paragraphs of arousal (the "hook," as journalism professors like to call it). His arousal efforts consisted of references to such nonscientific things as baseball, Mickey Mouse, architecture, opera, painting . . . all things that left the reader thinking, "Wow, this is interesting, but what does it have to do with evolution?"

That question, that arousal of interest, was his entry point to descend into the more sterile and less humanized world of science.

Why Scientists Need Artists

Gould led me to a minor revelation: that science, in itself, ain't real interesting to the broad audience. It simply isn't enough for the general public—it's too cold, too complex, too informational. It needs to be partnered with a more humanized element. This is why scientists need artists.

The typical cynical scientist looks at the work of an artist—some sort of crazed painting or dance routine—and chuckles, like Butthead, "That seems kinda dumb." But the work of art arouses people. It reaches down into those lower organs. Art stirs the heart, the gut, and even the loins. It motivates people. And that motivation can lead people to want to engage their brains. Which is when the scientist can go to work.

Arouse and fulfill. Supremely profound. Let no communicator fail to appreciate this partnership of elements. And let no individual fall victim to science-think, which in these terms turns out to be the mistake of believing that the formula is "fulfill and fulfill."

Heresy Warning: Film Is Not a Very Effective Educational Medium

So now the question becomes "How do you arouse your audience?" And this is where film comes in. It has enormous communication power. But before I talk about what it is good for, let me begin with how it has traditionally been misused in education.

My feelings about this go all the way back to 1967, to my seventh grade science class at Hocker Grove Elementary School in Shawnee, Kansas, where I was forced to watch the clickety-clacking, clattering, flickering educational science films on everything from the life cycle of the grubworm to human reproduction.

That was where I first developed my bitter hatred of boring "educational" films and my intuitive belief that film, by itself, just isn't a very effective way to teach students in the long run. Yet it took me the better part of a lifetime to move this gut instinct up to my head in order to come up with some logic to the premise.

The idea is that film is not a very effective educational medium but is indeed an incredibly powerful motivational medium. Let me start with some history.

Another Heresy Warning: Educational Technology Has Always Been Oversold

It's pretty much a rule that every piece of educational technology developed has been oversold. Every innovation, whether film, computers, or the Internet, is introduced as a panacea for the difficulties of getting students to learn.

The overall pattern has been well documented by Stanford University professor Larry Cuban. In *Oversold and Underused*, he examines the problem of computers in the classroom, and in *Teachers and Machines: The Classroom Use of Technology Since 1920*, he provides the definitive skeptical analysis of film's role in teaching. In the latter book, he offers up this wonderful quote from one of the inventors of cinema, Thomas Edison, who in 1922 said:

> I believe that the motion picture is destined to revolutionize our educational system and that in a few years it will supplant largely, if not entirely, the use of textbooks. I should say that on the average we get about two percent efficiency out of schoolbooks as they are written today. The education of the future, as I see it, will be conducted through the medium of the motion picture . . . where it should be possible to obtain one hundred percent efficiency.

Sounds like Edison envisioned a sort of *Clockwork Orange* future for education, where children show up at 8:00 a.m., sit in recliner chairs, have their eyelids propped open, and then watch movies until 3:00 p.m.

In fact, Cuban tells a story from the 1970s in which this basic idea was attempted in American Samoa. Being in such a far-flung location, the schools could recruit only limited numbers of qualified teachers. So they tried an experiment. Videotapes of some of the best teachers in the United States were made and shipped to the Samoan school district, where the students were forced to live out Edison's dream of watching the videos all day long. Within a few weeks the students rebelled and threw a television set out a window, and the program was ended. You can force-feed only so much media to people.

Though Edison's prediction that "books will soon be obsolete in the schools" still hasn't happened, his heart would probably be warmed to see the Internet playing the role he imagined for film.

Nevertheless, there exists this term "educational film," which I have concluded is largely a contradiction in terms. For starters, film is not effective for education because education revolves around one key trait—*inculcation*—the repetition of information as the brain creates the proper structure to retain it over time (do you need me to repeat this point?). We all recall having to repeat after the teacher in grade school. And much of the reason I learned so much in my acting class is that the teacher repeated the basic principles over and over and over again. But this is anathema to film. And even to storytelling.

At USC, this became glaringly obvious. It's almost a blanket rule that you get to use a piece of film only once unless you're doing some sort of memory scene later or a dream sequence, or maybe making a movie like *Groundhog Day* or *Run Lola Run*. But even then it gets tiresome and even angering to watch the same scene repeatedly.

In genuine education, it is essential to stop periodically and repeat all that's been covered and to repeatedly work and rework the material in different contexts. This is why education can get so boring and is a more active process than watching a film. The bottom line is pretty much that you get what you pay for when it comes to education.

So does this mean that films should never be shown in an educational setting? Of course not. It doesn't mean that at all. What it means is that it is essential for educators to know what they are dealing with when using the medium of film for education. And what they are dealing with is this . . .

Final (Good) Warning: Film Is an Excellent Motivational Medium

And that, very simply, is the truth. When *Top Gun* came out in 1986, one of my friends from high school saw it ten times in a single week, and then, massively motivated, he joined the air force—along with a lot of other instantly motivated young men. The U.S. Air Force Academy saw a measurable jump in enrollments.

Jurassic Park did the same thing for paleontology in the 1990s. And neither of those movies provided any statistics or logical arguments on why people should choose those careers. They provided motivation, not through the head but through the lower organs, by telling a good story with plenty of humor, excitement, and emotion.

There are countless examples of how movies, television shows, documentaries, and the stories they tell have motivated people in their career paths. But that doesn't mean films are good for conveying information and actually educating.

Let me tell you of a little exercise we did in one class in film school. An instructor showed two corporate training videos about how to run a drilling machine used in a manufacturing plant. In the first video, a man stood beside the machine and pointed out all the major parts, how they worked, and all the other details needed to run the machine. The camera never moved, and it was all very clear and thorough, but dull.

In the second video, there was no boring man speaking. Instead, the camera moved past the equipment, zoomed in on parts, and swished over the top of the machine as a seductive female voice narrated with lively music in the background. And the lighting was downright romantic.

Broad Audiences Prefer Style over Substance

Everyone voted the second film as the most effective. But when you analyzed the scripts, the second film had only half the informational content of the first and not nearly enough information to show how to run the machine. The class had opted for style over substance. And that is the basic dynamic of film as a medium.

The broad audience is very visual. The basic rule for making a film for the broad audience is "Don't tell us; show us." And the converse describes how this audience learns via film—things that are shown are much more powerful than things that are told.

The classic illustration of this in presentation videos is for a man to look into the camera and ask you to do as he tells you. He says, "Touch your ear," as you see him touch his ear, then "Touch your nose," as you see him touch his nose. But then he says, "Touch your chin," as he touches not his chin but his cheek.

The vast majority of the audience in this exercise will follow what they see rather than what they hear, touching their cheek rather than their chin. That is the override of the audio channel by the visual channel. And it gives you an insight into how fickle the medium of film tends to be. Visuals are extremely powerful and are used to tell the story.

In fact, at USC film school, the faculty is so keenly aware of this that for our entire first year we were hardly allowed to use sound. All five of our first semester films had to be silent movies in which images told the story. No dialogue, no narrator, not even on-screen text.

Film Is a Visual Medium

"Film is a visual medium" was the phrase they pounded into our heads week after week. In many classes, the instructors would have us watch entire movies with the sound off, just studying the use of visual images. When you begin to add up all the different elements available to a filmmaker—visual expression, music, sound effects, narration, storytelling, and so many more—you begin to realize that a film is infinitely complex.

And this takes you back to my simple calculation. If a picture tells a thousand words, and there are thirty frames per second in video (twenty-four in film), just do the math. You get 1.8 million words per minute, or 108 million words per hour. A typical novel has about 100,000 words. So, presto, in an hour you're reading the equivalent of 1,080 books!

Okay, that's total nonsense, because most of those thirty frames in a second look virtually identical. But the point is, you are indeed being given a great deal of information when watching a film, and most of it you don't really perceive or comprehend.

In the 1960s it was called "subliminal seduction," and for a while everyone thought it was everywhere. The idea was that films were full of hidden frames that made you want to buy more popcorn or have sex. Eventually people realized that subliminal stuff is effective only sometimes, and it's hard to predict exactly when. But there's no denying it can sometimes be powerful.

This is what Thomas Edison didn't realize with his new invention. He thought film was an educational panacea, applicable to anything that needed to be taught. It would take decades of exploration for people to understand that film is incredibly powerful for certain things and terrible for others. Most instructors still don't know this. But I can give you an example of how it works in my specific field of science.

There are about thirty-five major groups of invertebrates in the world. These are the animals that lack backbones. Some groups are innately more interesting than others. Some of the major groups just look like worms and are relatively boring. Other groups include things like octopuses and squid, which are endlessly fascinating as they change their colors in a second and have eyes as complex as ours. Or giant crabs or lobsters or bombardier beetles or walking sticks—all kinds of amazing creatures.

When you teach the biology of invertebrates, you need for students to learn all of these groups, so in one lecture after another you go through them and do your best to make the worms interesting. It's not always easy, but that's what education involves—laying out the details in an organized, systematic manner, going back over the material, repeating the material the students find difficult or unclear. Back and forth, back and forth, creating the structure in the brain where it can all be retained.

Back to the Old Arouse and Fulfill

So if you want to make a so-called educational film about the invertebrate groups, you have two choices. Option one, you plod through all thirty-five groups one at a time, spending an equal amount of time on each group, telling about one worm group after another. By the time the students have gotten through their fifth worm group they are hating the film, hating the animals, desperately awaiting the group with the octopus, and mostly just wanting to get the entire experience over with. And then, every so often, you stop and go back and review all the groups to make sure everyone is up with you. And that is just too much for the students as they throw the movie projector out the window like the Samoans did.

The other option is to make a film that, instead of focusing on education (i.e., covering *all* the material), focuses on entertainment—telling only about the groups that are immediately fascinating and compelling. The film is light and breezy, highlights some of the more interesting features of the more interesting groups, and leaves the students wanting to know more. The film is only ten minutes long, and when it ends, the students are full of questions about how the cone snail manages to capture fast-moving fish or why the male isopod crustacean lives as a parasite on the female.

In other words, back to the old arouse and fulfill. Use the film to arouse—it's a mighty powerfully stimulating medium when used properly—and then step in to deliver the fulfillment. This is exactly the model we developed around *Flock of Dodos*. The film is a fairly light eighty-five-minute romp that touches on some elements of the evolution–intelligent design controversy. It arouses the interests of the audience but doesn't really fulfill them. Which is why we have held so many events in which the movie is followed by a panel discussion with several experts in evolution, theology, communications, and government. They are the fulfillment. And on the home DVD version we added a bonus feature—the ten most commonly asked questions in the panels, answered by several experts.

The Magic of Juxtaposition

As I've said here, film is simply not a very powerful literal medium. Sitting a bunch of experts down for interviews and having them talk directly to the camera is instantly boring to most viewers. Anything short of Elvis Presley or a serial killer and the mass audience tunes out. And just think of watching talking heads with the volume turned off. You'd have no idea what the film was about, meaning it could communicate only through sound, the weaker channel.

But film has the potential to be an incredibly powerful nonliteral medium. For instance, if in 1964 you had made a television commercial in which a man looked into the camera and said that Republican presidential nominee Barry Goldwater was dangerous, it would have had little, if any, effect on the election. But if you had produced an ad about Goldwater in which a scene of an innocent little child was followed immediately by footage of an atomic bomb explosion . . . well, put it this way. His opponents did. It ran only once and was enough to undermine Goldwater's entire public image.

That's the magic of *juxtaposition*—two unrelated images lined up against each other, producing something more powerful than their sum. My acting professors tried to teach us the concept by showing montages by Sergei Eisenstein and Luis Buñuel. It took over a decade for the idea to really percolate into my brain, but I've intuitively tried to use creative juxtaposition since I first got involved with filmmaking.

One of my first reasons to begin experimenting with film was that, by the late 1980s, I was bored, frustrated, and disappointed with most of the nature documentaries about coral reefs I had seen on television. I had spent years diving on some of the most eye-poppingly inspiring reefs in the world. My brain was still fresh with memories of the swimming pool–clear waters of the Great Barrier Reef, where you could see schools of sharks chasing fish a few hundred feet away. Those experiences left me exhilarated. Why couldn't the documentaries do the same?

Literalists would say, "They can't; they're just pieces of film. It's not the same thing as being there." Nonliteralists would say, "They could through juxtaposition, but they don't because the filmmakers just aren't very good."

I still believe in the latter—that film has infinite power. And guess what—that's not a falsifiable hypothesis, so you can't tell me I'm wrong!

One of my experiments in this direction, even before I went to film school, was my barnacle music video, *Barnacles Tell No Lies*. I thought barnacles were cool, and I wanted to convey this to a broader audience. The literal thing to do would have been to show footage of barnacles while a narrator said, "Barnacles are amazing creatures that have a fascinating ability to move their appendages rhythmically with the flow of the water." That would have been a case of *telling* the viewer that the creatures were amazing and fascinating.

Instead I opted to *show* the viewer by setting footage of the appendage movement to the rhythm of a jazz song and then throw in a little of the lower organ elements by having a sexy jazz singer with a seductive smile serenade the barnacles as the film reveals they also have the longest penis relative to body size of any animal. Suffice it to say, unlike most of my embarrassingly bad early film efforts, the barnacle movie has stood the test of time. Nearly two decades later, viewers still laugh and walk away impressed, realizing that barnacles are much more amazing and fascinating than they thought. And the reason why is the juxtaposing of serious science with silly and even sexual humor. It does something. It transcends. It's nonliteral. And it has the potential to be extremely powerful in mass communication.

And Now It's Time for a Story . . .

So, arouse and fulfill. I can't emphasize its importance enough.

And yet . . . there is another way to deliver information that is different from this two-step process. Instead of partitioning the arousal and the fulfillment, there is an age-old way to mix the two together into a single, endlessly digestible stew. It's called storytelling, and guess what? It remains today the most powerful means of mass communication. And, sad to report, scientists have some problems with it, as you will learn in the next chapter.

THEREFORE: UPDATE FOR THE SECOND EDITION

Eternal Sunshine of the Literal Mind

Ah, literal-mindedness. It's never going to change. You see it everywhere. It's the bane of bad communication. But, for better or worse, one trend has emerged since I wrote this chapter that could improve things if done right. It's that everyone nowadays feels the need to do a better job of "telling their story." Storytelling is the counterbalancing force to literal-mindedness, so there is cause for hope.

Storytelling has become the hot trend. On Twitter, the number one self-descriptor in profiles is "storyteller." TED Talks are packed with discourses on "the power of storytelling" (actually, I gave a TEDMED talk in 2013 on the power of the ABT narrative template, so I guess I'm part of the trend), and a boss at Deloitte brought me in to run a workshop with the company's executives because "they are being told constantly that they need to do a better job of telling their story."

So how is storytelling less literal-minded? Let me start by showing you exactly what I mean by literal-minded in the real world.

The Prearoused Surf/Climate Crowd

I talk a lot about how "focused on their own world" scientists can get, but the fact is it can happen in any profession, hobby, or sport. Before this book came out, my older brother, a lawyer, read a draft and said, "You could write the same book for the legal profession titled 'Don't Be Such a Lawyer.'" The same is true for accountants, politicians, mechanics—pretty much every profession potentially involving heavy doses of information.

Here's my favorite parallel example from the nonscience world.

I'm a mad keen surfer. I'm at my happiest when I can surf three to five times a week. I do stuff with the Surfrider Foundation, my favorite environmental group. Years ago I spoke about ocean conservation at an event for a Surfrider chapter in Southern California along with LeRoy Grannis, one of the greatest surf photographers ever. He was eighty-seven at the time. I talked about "shifting baselines in the surf community," and then he talked about . . . well, he didn't really talk much. It was at a big, fancy house, there were about fifty people present—mostly young surf dudes and their girlfriends. All LeRoy did was show three eighty-slide carousels of his old 35 mm slides.

Half of his audience was prearoused—the surf dudes. They didn't need the power of narrative. They sat in dead, reverential silence (think of a room of twenty-five Jeff Spicolis from *Fast Times at Ridgemont High* staring mesmerized at the screen, letting out the occasional "Awwwwwesome") as he literally clicked through one slide after another—probably five seconds on each slide—twelve slides a minute—about ten minutes per carousel. The young dudes had erections just looking at the amazing photos of the world's greatest surfers shredding down the faces of the world's gnarliest waves.

But the girlfriends, one by one, made their way out to the kitchen. After a while you could hear the dull roar of them chatting away back there as LeRoy was holding church.

That's your prime example of the rare instance in which you don't need the power of narrative. It's where you've got an audience so thoroughly aroused from the first moment that they mostly don't even want to have a story get in the way of their soaking in every moment of every photograph. Basically, "Shut up with telling me about Kelly Slater's childhood in Florida—you're distracting me from focusing on how his trailing foot is only half on the tail pad as he's snapping that bottom turn."

In the climate world, the surfers are the climate community, LeRoy is Al Gore, the girlfriends are the rest of humanity—bored stiff and out in the kitchen, not listening. So the literal-minded crowd is "the choir" and is easy to reach. The way you reach the others is through the three forces of narrative.

The Three Forces of Narrative

When I wrote this chapter for the first edition, I didn't know about the three forces of narrative (agreement, contradiction, consequence). Now I do, and I see that the forces explain pretty much everything.

A story, properly told, starts with agreement—you lay out a few things we can all agree on. This is where so many scientists disconnect. They never want to have to tell you something that "everybody already knows." They fear it will make them look stupid. They are literal-minded, meaning "We're here to talk about a problem; let's not waste any time—let's do the more literal thing and jump right to the problem."

You could call it cutting to the chase. But it means skipping that first element of agreement, cutting right to the contradiction before we even know the basic setup. For narrative to work best, there has to be the setup. That's how you open up the channels of communication before they get clogged with contradiction.

I get to see this dynamic in detail with a major drug development biotech company with which we run our Story Circles Narrative Training program. It is a corporate science institution that maintains a fiercely information-intensive environment. I get warned by the bosses that there is zero tolerance for "fluff" in talks. For research scientists, the idea of telling the audience something that everyone already knows is not just their worst fear; it's the sort of thing that will cost them credibility among their peers and, ultimately, their job.

The result is talks that start with a slide containing five graphs. Seriously. No setup, just dive right in. They live in fear of boring anyone, so they cut to beyond the chase—all the way past the chase to flying over the cliff. And yet the bosses know the talks have a problem with clarity—especially when the room is full of VPs who are not in the thick of the subject matter. Which is why they bring me in.

It's been a slow, painful struggle, but I'm getting at least a few of them to see the value of being a tiny bit less literal-minded. Instead of cutting to the chase, take some time with the agreement element. In fact, in a slightly different context, let me show you more clearly how this investment in agreement pays off by making it possible to reach the massive non-choir part of the public.

How the Big Boys Do Narrative: HBO's *Real Sports*

People often ask me what the best science documentary show on television is. My answer is that it would be HBO's *Real Sports with Bryant Gumbel* if they ever decided to do science. To make this point, let me tell you two stories about the death of coral reefs.

I was invited to take part in a post-screening panel discussion of the 2017 documentary feature film *Chasing Coral*. I watched the first ten minutes of the movie on Netflix and said, "No, thanks; you wouldn't want me there." I'll explain why in a bit.

Just two days later, as I was still feeling bad about declining the invitation, *Real Sports with Bryant Gumbel* ran a segment on the dying of Australia's Great Barrier Reef that confirmed my instincts. The piece was so cleanly done—so completely embodying all the rules of narrative that I teach in Story Circles— that I transcribed the entire twelve-minute segment and wrote a detailed blog post about it, pointing out all the skillful narrative attributes.

Now here's the important and relevant part. Effective communication begins and ends with how you open a story. *Chasing Coral* did a jumbled-up job with its opening material, while *Real Sports* did nearly perfect work.

What I mean by the opening is the first of the three basic acts that make up a story. The first act has an ideal form that is pure ABT (And, But, Therefore) structure. It's meant to be the largely non-narrative "ordinary world" described by Joseph Campbell, where very little goes on. It's the stage of agreement before the contradiction begins, which is quickly followed by consequence. Think of it as the start of a murder mystery, where everything seems normal in a little town before a dead body is discovered. It's basically "All is quiet AND everyone is happy, BUT then a dead body is found; THEREFORE, an investigation begins"—end of first act.

When you look at what the professionals at *Real Sports* did (and they are true professionals, having won more than twenty Emmys and two Peabody Awards), you'll see they followed exactly that form. They open their segment with three minutes of pure agreement. We follow Bryant Gumbel scuba diving on a beautiful coral reef. He talks to the guide about all the amazing wildlife they see, they marvel at all the diversity, there's nothing controversial; no real story is being told until we hit the three-minute mark, and then Bryant says, "But there's a problem . . ."

Then you look at *Chasing Coral*, and what do you see? Within the first five minutes—less than 5 percent of the show—there's the host telling about his childhood, about working as an ad agent; there's a comment on beautiful reefs today, there's mention of one problem (dying coral reefs), then a second problem (unable to communicate it), and then we're visiting a dead reef before we really know what a healthy reef looks like. It's a hodgepodge of agreement, contradiction, and consequence repeatedly, all over the place, before we've even started a clear journey.

Complexity: The Bane of the Amateur Mind

The comparison of simplicity from the professional with overcomplication from the less experienced filmmaker makes me think of the most memorable class experiment we did in film school at USC. The instructors broke us into groups, gave us one page from the screenplay for the iconic movie *Chinatown*, and told us to make a list of all the shots we would use to film that one page of material.

It was the scene of the young Mexican boy riding his horse in the riverbed past Jack Nicholson. For my group's list, we had a wide shot of the riverbed, a close-up of the child, a close-up of Jack, a close-up of the horses' hooves, a Steadicam shot of the horse and child riding along, a crane shot rising up from behind Jack . . . on and on. So many cool shots!

Each group wrote its shot lists on the board. These ranged from about fifteen to twenty shots. Then we were shown the actual scene in the movie. Guess what.

Three shots.

That's all. Very simple, very clean, very elegant. It was such a dramatic demonstration of what the mind of an amateur does, which is to overthink everything (National Academy of Sciences, this means you).

We all felt so stupid. And that's the difference between the overly busy and complicated amateur storytelling of *Chasing Coral* versus the clean, sleek, simple professional storytelling of the *Real Sports* team.

One more tag to this anecdote: a week after my blog post I received an e-mail from Chapman Downes, producer of the HBO *Real Sports* segment. Someone had forwarded him my post. He thanked me for the praise, and we ended up having a great hour-long phone call in which I drilled him for details on how and why they are such great storytellers at *Real Sports*. I can tell you the secret to their success in two words: narrative intuition.

The show has been around since 1995. The main producers have all been there for at least fifteen years. They have the sort of deep intuitive feel for narrative that accounts for their success. And the one thing I can guarantee you is that you will never see their work suffer from literal-mindedness—they always have a great story to tell.

CHAPTER 3

Don't Be Such a Poor Storyteller

I want to share with you the single most humiliating public experience of my life. In the spring of 1990, Spike Lee's movie Do the Right Thing *was hitting America, I was a professor at the University of New Hampshire, and suddenly Spike was on campus for a simple event called "Open Mike with Spike." More than a thousand students packed a huge room in the student union just to ask him questions from two standing microphones. I decided to take a shot. I had just made my first foray to Hollywood with a screenplay, so I began telling him about my trip, my meeting at Columbia Pictures studios, what the executives I met with said, and a whole lot of other things—but something strange started happening about five minutes into my comment/question. I began hearing this reverberating, echoing sound that was bouncing around the massive auditorium. I couldn't quite make it out at first except I realized it was voices—a lot of voices—student voices—hundreds of them—a chorus—and then I finally paused my speech for a moment and heard what they were chanting. "Get to the point, get to the point, get to the point!" A wave of terror swept over me. I looked back at Spike and finished my speech by quickly blurting out, "So, like, what's up with that?" Then I put my tail between my legs and walked, head down, to the back of the hall. It turned out the event was being broadcast live on the student radio station. The next day a student stopped me in the hallway of the biology department and said, "Professor Olson, was that you last night asking that half-hour question?"*

You Bore Me

I used to have a German girlfriend. She was very funny—she came from Bavaria, where they love to laugh. I also used to be invited to go on trips with Harvard University alumni as a "guest naturalist," meaning I would help explain what the old folks were seeing in nature during trips to Norway, Antarctica, Australia, and Central America. I took this girlfriend with me on several trips. She would listen to me talk and talk and talk to the old folks, and finally, by the end of each day, she would have had enough. So her favorite thing to do in the evenings was, when I was done talking, to look deeply, romantically, lovingly into my eyes and say in a soft and seductive Germanic voice . . . "You bore me."

Which was true. I bore myself sometimes. I learned the art of boredom from my father. He was a military historian, and we were his pupils—my two brothers, two sisters, and I—at the dinner table. He served in Vietnam as a troop advisor in the early 1960s, and he felt a deep need for us all to understand the depth and complexity of the Vietnam problem. But the lectures on Vietnam weren't just about what was going on there at the moment. Oh, no. That would have been too simple and relevant. No, his lectures had to begin at the beginning, back before the American involvement, before the French involvement, back . . . oh, I don't know, maybe in the Paleozoic era or something. He would drone on and on for hours, not telling a story, just ambling about, relaying a stream of consciousness made up of all the disconnected factoids and tidbits floating around in his head. And we were like that "get to the point" audience. (How in the world could I have ever made the same mistake in public? Had to be a genetic element at work.)

Here's a big surprise: I grew up to be a scientist. And, guess what . . .

Scientists Are Poor Storytellers

Do you really need proof of this? If you do, go visit a research laboratory, walk into any lab, and ask the guy with the thickest glasses, "So what are you studying here?" Then take a seat, put your elbow on the table, chin in palm, and settle in for a half-hour ramble. How do I know this? Because I was one of those guys when I was a postdoctoral fellow at the Australian Institute of Marine Science. Tourists would ask about my automated underwater starfish larvae growth chambers and I would unleash a long-winded discourse that I felt wasn't finished until I had put out the fires of their curiosity. "D'ya wanna know more?" I'd proudly ask their backsides as they fled out the door.

It's a problem. Another girlfriend developed an affectionate nickname for me, "Chief Longwind," which she would abbreviate when I'd get going on something and just say, "That's enough for tonight, Chief."

If you take a look back at those wonderful Stephen Jay Gould essays in *Natural History* magazine, what you'll see, as I mentioned, is a clear partitioning of the opening few paragraphs, providing the arousal, and then the following few pages, giving the fulfillment. This worked to a point, but the truth is that some of the opening hooks did make one wonder, "What exactly is this going to have to do with science?" Then, especially in his later years, he slipped into overstaying his welcome in the fulfillment parts. Some essays had page after page of minutiae about taxonomists and natural history. He lost even me. The arousal bit can take you only so far with the reader.

So the arouse-and-fulfill strategy has its limitations. For all you scientists out there, it's kind of like the surface area to volume function in limiting organism size—you eventually reach a point where there's not enough surface area for gas exchange and the organism can't get any larger. At that point, the organism has to have a circulatory system. In other words, there has to be a different way of doing things. For communication, that different way, beyond the simple arouse-and-fulfill model, is storytelling.

It is an enormously powerful means of communication. With good storytelling you end up both arousing and fulfilling at the same time, which allows you to sustain interest over much larger amounts of material.

Storytelling is equal parts art and science. And there we have it already. Like all these other things we've been discussing, it's made up of two parts. One is more objective, the other more subjective.

The first part, story structure (or just story), is the objective part of telling a story. There is a science to story structure. It is something that can be taught and analyzed. The major studios in Hollywood have story departments, with story editors and story analysts. At film schools there are countless courses in story writing in which the fundamental components of telling a story are taught.

Most screenplays, for starters, have an incredibly formulaic structure that is nailed down almost to the page. A standard screenplay is about 120 pages and has three acts that, in round numbers, are about 30, 60, and 30 pages in length, and each page roughly equals a minute of screen time. Within these three acts there are a number of points of structure, such as the "first plot point," which by convention generally occurs somewhere between pages 23 and 28. This is the point where the calm and quiet world of the opening act is suddenly overturned by the major "inciting incident"—the kidnapping, the murder, the declaration of war. Then there is a "midpoint" somewhere around the midpoint of the script (big surprise) . . . On and on, lots and lots of structure, which allows script analysts to determine if the formulas are being followed and, if not, to bring in a "script doctor" to fix things.

Story structure brings with it a great deal of seemingly objective rules and conformity—the sorts of things that make scientists very happy and content.

But there's a second element called character, and guess what? It's much, much more subjective. Where story is toward the science end of the spectrum, character is more toward the art end. Character is the way the actors talk and dress and walk and pose and laugh and all those things that end up being what people imitate years later when they talk about their favorite movies.

One of my favorite quotes from film school is from an interview we read with Rex Ingram, director of the classic 1921 silent movie *The Four Horsemen of the Apocalypse*, in which he said, "Nine times out of ten it's character that people remember from their favorite movies rather than story." And that's pretty darn true.

Think of your favorite movie. Maybe it's *Casablanca*? You remember Humphrey Bogart as Rick Blaine, the bitter, cynical American expatriate, and all his famous lines about rounding up the usual suspects, beautiful friendships, and playing piano songs. What most people don't think of is the story point where Rick double-crosses Renault. You think of the characters that inhabit your favorite movies—Rocky, Rhett Butler, Dorothy Gale, Forrest Gump—not the story lines.

Character is so much more powerful and deep and complex, but it is also very elusive, hard to teach, hard to analyze. Sound familiar? It's the same divide as substance and style. And so you can imagine that if a scientist were to become a screenwriter, he or she would probably be naturally more drawn to story than to character—telling precisely crafted, intricate stories in which all the facts add up, but . . . the characters are dull.

But story is also incredibly important. And even though character is such a powerful and memorable part of it, it is through the properly structured story that the true magic emerges. From film school to the present I have been slowly and surely learning this, the hard way, through trial and error.

Entering film school at the University of Southern California, a lot of us thought we were great and gifted directors who very soon would be directing massive-budget movies, making them work like precision machines without even breaking a sweat. Three years later we were all pretty much wrecks, our self-confidence in shambles.

Film school will do that to you as you learn of the infinite complexities of film (remember all those elements I itemized earlier) and find out that telling a clear and simple story is a true art. The variety of elements alone allows enormous complexity. Then consider the sheer and total insanity of actors (just think of my crazy acting teacher). You begin to realize that instead of running a precision machine, you're trying to drive an old jalopy with a loose steering wheel and mud all over the windshield. Directing a movie is not easy.

In our first semester we were taught a simple old adage, "If it ain't on the page, it ain't on the stage," which means that if you haven't invested immense time and energy in the writing of a really good script, which gives everyone involved with the movie a clear picture of the finished product, you probably aren't going to end up with a very good movie.

I thought this was nonsense at first. I was a brilliantly talented filmmaker who had a crystal clear vision of the films I wanted to make. I didn't need to slave over some tedious script. I knew, deep in my heart, that I could take any script, no matter how ramshackle, dull, and pointless, and, with my actors and my camera positions, craft it into a masterpiece.

The only thing I didn't know was that I was really naïve. And would have to learn things the hard way. Through the school of hard poundings on the head.

When I finished film school I was forty-two years old—already practically retirement age in Hollywood—and despite the award-winning musical comedy I had directed in film school I was offered a big fat nothing when it came time to find work. (It was 1996, and I was told by every agent and manager I met that musicals were a thing of the 1950s, dead and over, never to be seen again, despite *Moulin Rouge!*, *Chicago*, *Evita*, and a stack of other hugely popular musicals that would emerge a few years later—ah, Hollywood.)

I wanted to direct comedy movies. In the end, the only opportunity to direct a feature comedy I could find involved a couple of young actors who wrote a not-so-great script, cast themselves as the leads, found a private investor, and hired me to direct it. I threw my massive ego into the project, determined to show I could turn any sow's ear into solid gold—and I failed. We ended up with a movie that, while fun, had not-so-great acting, didn't tell a good story, and had nobody wanting to buy and distribute it.

The whole process was deeply painful. It drove me down, by 1999, to the deepest, bleakest depths of Hollywood despair. All of the big agents and managers who loved my musical (even though they told me the genre was dead, they appreciated what I had done), and who were still trying to think of ways to get me work, took one look at the movie and it was over. In an instant. That's what they do in Hollywood: you get your one shot. Those guys all shook their heads and basically said to themselves, "Ah, just like all the other schmucks. We knew he couldn't direct."

It would take seven years for me to get over the trauma of that failed movie. But when I finally did, it occurred through a near religious experience in the making of *Flock of Dodos: The Evolution–Intelligent Design Circus*, which brought me around to finally understanding, through much pain and suffering, why there is so much focus on storytelling in Hollywood.

It was the most powerful experience of my filmmaking career, and it reveals so much that I must now go through it in excruciating detail—though I'll try to get to the point fairly quickly so you don't start chanting at me.

Forging the Story of *Flock of Dodos*

The point of this section is that you have to have a story. There. I got to the point. Are you happy now?

As I mentioned earlier, in 2005 I read an article in the *New Yorker* titled "Devolution: Why Intelligent Design Isn't" by H. Allen Orr, which prompted me almost instantaneously to start filming *Flock of Dodos*. Yielding to my improv and Meisner training, I quickly assembled a crew, flew to Kansas, and spent a week filming interviews. Then we went to the East Coast for another week of interviews, and before I knew it I was sitting in our editing suite in Los Angeles showing these interviews to friends and listening to their enthusiastic responses.

What they all said was the same thing, over and over: "This is incredible raw material. Now if you can just put it together into a story, you'll have a good film."

Yeah. Very simple. Like standing in your living room looking down at the floor, where hundreds of unassembled pieces of a gargantuan Ikea combination desk/dresser/chopping table are laid out, but you have no instructions or picture of the finished product. Just a bunch of friends admiring the beauty of the unassembled pieces and saying, "We know you can do it—we can sense you've got something great here."

And the extension of that is, "However, if you fail to put these pieces together in a way that works, we're going to write you off as a total loser and make you feel guilty for having taken up the time of the good people you interviewed."

The pressure began to build as we started editing. And now I want to tell you how this film came together on a weekly basis so you can, I hope, appreciate the magic and power of storytelling as I did.

In the first week of editing I created what is called an "assembly cut," which means I took all the interesting pieces of interviews and interesting scenery and spliced them together into about a three-and-a-half-hour cut, which I showed to only one person, my good friend and trusted longtime producer, Ty Carlisle.

He said, "Okay, nice start. You've clearly got the goods; now get to work crafting a story."

It was like a giant piece of marble, ready to be sculpted into the Venus de Milo. But for now it was just a giant, shapeless lump.

The next week I started putting together sequences. I made up a bunch of rules for myself, like "The story needs to start with the quirky little tidbit about my mother being neighbors with the big lawyer for intelligent design in Kansas, John Calvert," and "It needs to work its way to the grand synthesis of what evolution is and why it's so important to teach," and others that I made up as I went along.

That Friday I called together the six or seven folks working with me—my sound editor, animator, producer, cameraman, and so on, for a viewing in our office. This cut was about two and a half hours. When it ended and I turned up the lights, everyone seemed exhausted and had pages and pages of notes. We began talking about the movie, and though everyone echoed what Ty had said the week before—that the raw materials were there—nobody was smiling.

They began going through their notes, and disagreements broke out. One person said, "You need to open with the Dover trial, since that's what's in the news right now," and another said, "No, you need to finish with the Dover trial; it brings us up to date with current events." On and on, for an hour of nearly complete disagreement. Everyone's suggestions were huge in scope—like "Move this entire section to the front"—and nobody felt particularly confident in their suggestions.

I went to work on the next cut, this time opening with a description of the big, bad Discovery Institute, which led into the topic of intelligent design and then . . . on and on. That third week's version was down to about two hours, but when I flipped on the lights at the end of the viewing, the faces looked even grumpier and everyone still basically disagreed.

Worst of all, I began to split off from the rest of the group. I began to say, "I'm actually liking this cut—it's starting to work pretty well for me," while the others were saying, "It still doesn't tell a good story. Nobody other than professors will want to watch this." Our disagreements were intensifying, and I was starting to take it personally—which led to the big blowout at the end of week four.

"Finding the story" for "Flock of Dodos"

Week 1	Assembly Cut (3.5 hours)	Academics
Week 2	Calvert/Muffy Moose, Disc. Inst., Dover, I.D., Kansas, Evolution	Teachers
Week 3	Disc. Inst., Evolution, Kansas, Muffy Moose/Calvert, Dover	Teachers
Week 4	Dover., I.D., Disc. Inst., Evolution Kansas, Muffy Moose, Calvert	Teachers

Week 5	PROLOGUE	Muffy Moose
	ACT I	Evolution & Intelligent Design Journey: Off to Kansas
	ACT II	Kansas in search of Calvert Confront the Dragon
	ACT III	The Real Dragon: Discovery Inst. Resolution: Information Wars

Figure 3-1. The "evolution" of the story line for the documentary feature *Flock of Dodos* during the first five weeks of editing.

When the lights came up at the end of the week four viewing, the gloves came off. I had a smile on my face and said, "Looks like it's about there." No one else was smiling. Everyone said, "Sorry, but it's just not a story. You've got a few good sequences here, but it doesn't add up, doesn't build toward anything. It's . . . boring."

Of course, those are fighting words, which is precisely what happened. I ended up snapping at everyone, telling them they were blind, that it was a great movie and it was almost done. It *had* to be done because we were running out of money. But they all held their ground and withheld their praise (my friends are tough). And I erupted, shouting at them, telling them to get out of the editing suite, that they sucked and didn't know what they were talking about.

When they all left, I closed the door and sat in front of the computer, and the darkness began to settle in. I stared at the clips on the screen and began to realize we had $200,000 tied up in something that my crew was telling me would never make it out of classroom viewings. This couldn't happen. So I plunged headfirst into the abyss.

I called Ty and told him to take the next three days off. I went home and got a few days' changes of clothes and then came back and went to work. I did the standard movie-writing thing of filling out index cards for every scene and placing them all over the floor. And I began searching for "the structure" of a possible story.

I ordered food delivered. I slept on the couch. I cut and recut the scenes. And, slowly but surely, a very simple (and in retrospect obvious) story began to emerge.

It was the story of a man who sets out on a journey to save a damsel in distress. He must protect her from the dragon that lives next door. But when he finally confronts the dragon, it turns out to be a teddy bear. He realizes the real threat is not the dragon but an evil empire, and in the third act he goes in search of it.

That ended up being the more or less "mythic" structure beneath the story I told for *Flock of Dodos*. I was the "man," the "damsel" was my mother, her "homeland" was Kansas, the "dragon" was her neighbor, the lawyer for intelligent design, and the "evil empire" was the Discovery Institute in Seattle.

Somewhere around Wednesday night I finally hit on these revelations and pulled together the rough pieces, and Friday afternoon, when I unlocked the door and allowed everyone back inside to view what I had assembled . . . a miracle happened.

When I turned up the lights at the end of the viewing, there were smiles all around. People said, "That was fun," and "That blew by in what seemed like about thirty minutes" (it was still nearly two hours), and, most important, "You've finally got a story."

Their notes were now minuscule. Instead of suggestions for moving huge blocks of material from one section to another, the suggestions were things like "You should add a few more seconds of Dr. Steve Case and maybe make a graphic to illustrate what he's talking about," "We need more of your mother," and things of that sort.

Never again was there a major frown of frustration or boredom from viewers, whether during an editing screening or at a public event. The simple structured story has carried the film through hundreds of public screenings with all sizes of audience. Bring the lights down, tell everyone a simple story, and they will allow you to get away with all sorts of things. It is truly magic.

And that's how you get beyond the arouse-and-fulfill dictum. Keep the story going and you can keep the flow of information going . . . forever. That's what a good television series is—an ongoing story, week after week, feeding you information about the characters and story.

But There's a Catch: You Have to Suspend Disbelief

So now we know how to convey information in a wonderfully enjoyable and painless way through the telling of a story. It would seem that if scientists were interested in communicating at all, they would use storytelling at every opportunity. But there's a catch.

For an audience to enjoy a story, they have to take part in an exercise of trust known as the "suspension of disbelief." The poet Samuel Taylor Coleridge, who wrote *The Rime of the Ancient Mariner*, coined the term in his *Biographia Literaria* when he referred to "that willing suspension of disbelief for the moment which constitutes poetic faith."

The audience has to be willing to believe the storyteller at every turn and not bog the process down by asking themselves, "Do I really believe this could happen?" If you have to ask yourself that, you can't enjoy the story. It's a fundamental rule of storytelling, and it's where scientists get left out of the picture because their job is to question *everything*.

This is what scientists do for a living: they are trained *not* to take the bait. When you give a scientist a paper to read, instead of being your typical rube and believing every word simply because it's in print, he or she will question the premise, question the assumption, demand to see data, demand that you cite your sources—scientists just aren't gonna go for a ride in a car until they've kicked the tires and looked under the hood. This is why the phrase "Scientists agree . . ." actually means something. But it comes at a price—actually, a couple of prices. The price of storytelling, here, and the price of "likeability," which I'll discuss in the next chapter.

The refusal of scientists to suspend disbelief occurred with my film *Sizzle: A Global Warming Comedy*. The film is a "mockumentary," mixing the reality of my being a scientist-turned-filmmaker with the fictitious premise of my trying to make a documentary about global warming that runs into countless problems. The mix of genres ends up splitting the scientists out of the audience.

This is not to say that scientists can't enjoy plenty of stories. But still, I promise you, they simply do not enjoy them as much as the general public. They view themselves as the "designated drivers" of the storytelling audience. While everyone gets drunk on entertainment, the scientist maintains a certain level of sobriety, always keeping an eye on the facts.

I remember seeing scientist Carl Sagan on *The Tonight Show* with Johnny Carson in 1977, talking about the new science fiction movie called *Star Wars*. He agreed that the film was wonderfully fun, but he said he was still disappointed at tiny details they didn't have straight, like Han Solo using the term "parsec" as a unit of time when it's actually a unit of distance.

And, yes, now you're thinking, "Well, that happens for anyone—if I go see a film shot in my hometown, the Bronx, and there are snow-covered mountains in the distance (as there were in Jackie Chan's campy *Rumble in the Bronx*, shot in Vancouver), I'll have the same problem enjoying the film." Yes, but it's different for scientists because this mind-set is such a fundamental way of life in the profession of science—it is applied to *everything*.

Archplot, Miniplot, and Antiplot

It's worth taking a minute here to delve a tiny bit further into story structure and why it is such a fundamental part of communication. The telling of stories is how we come to understand our lives.

One of the best books written about it is *Story*, by Robert McKee. McKee identifies three types of plot and describes their structure by using a triangle (see figure 3-2). At the top of the triangle is the classic blockbuster movie story line, which has mythic structure underlying it. He calls this "archplot" (pronounced arc-plot).

Archplot produces what McKee calls "classical design," meaning all the standard things we think of—a hero sets out on a journey to combat the forces of evil, is faced with challenges, has lots of ups and downs, and eventually succeeds,

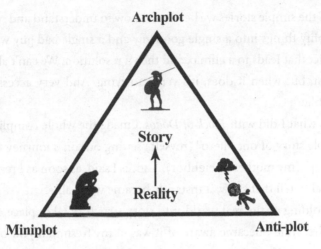

Figure 3-2. Triangle of story plots adapted from Robert McKee's *Story*.

concluding the story with a happy ending. This structure includes everything from *Star Wars* to *Rambo*.

At the base of the triangle are two types of movies that don't do those things. Miniplot is pretty much the opposite of archplot—there might not be a single hero, the struggle might not be against bad guys but instead might be within the hero's head, there might be many enemies, and the story concludes with an ending that can be vague and unresolved—an "open" ending. These are smaller, more artsy movies, like *Tender Mercies* and *Paris, Texas*.

Antiplot is the other extreme, where plot is simply thrown out the window—no interest, care, or concern for telling a story. Events jump around randomly, things happen for no particular reason (including coincidence), and not much adds up logically. This includes crazy movies like *Monty Python and the Holy Grail* and experimental films like *Meshes of the Afternoon*.

So much of our daily lives consists of having real-world experiences that are somewhere near the base of the triangle—a long way from archplot, and maybe even in the realm of antiplot—just a bunch of random events. But the way we make sense of events is by editing, trimming, rearranging, and massaging the information in an effort to slowly move it toward archplot. We try to make it into one of the simple stories we best know how to understand and relate to. We try to simplify things into a single good guy and a single bad guy with a single clear conflict that leads to a climax and then a resolution. We can't always make this happen, but when it does, it's very satisfying. And very accessible to the general public.

This is what I did with *Flock of Dodos*. I made the whole complicated issue into a simple story of one "hero" (myself) setting out on a journey to confront one "bad guy" (my mother's neighbor). And, as I said, as soon as I rearranged all our material to tell that story, it instantly became watchable.

The molding of the real world into story structure takes place every day. I think the first time I became aware of it was in my freshman year of college at the University of Kansas, when I was living in the Sigma Nu fraternity house. Night after night we would "go out drinking" at the bars, moderately interesting things would happen, and then, the next morning, as all the guys would awaken with hangovers, the stories would be told. And lo and behold, an evening's worth of random events would, in the minds of the better storytellers, emerge as something much closer to archplot—for example, one of our guys at the bar "who was just minding his own business" (the hero of our story) had a beer splashed on him by a jerk (the antagonist), who then called him a name (crisis), and onward as a not-that-simple evening is reworked into a simple, fun story that everyone can follow. The fact that our hero was also a jerk and was hardly splashed with the beer and didn't really feel that challenged—those are all details that got left out in the interest of telling a better story.

The key point is the fundamental movement from miniplot or antiplot to archplot as a means of reaching a broader audience. And now it is time for all of us who are scientists to brace ourselves and come to realize that we are no better than the rest of the human race when it comes to communicating our science, because . . .

Scientists Are F****s?

Okay. Sorry about that heading (the word is "frauds"), but I didn't come up with it. It comes from a Nobel laureate who, judging from his writings, was a very cool fellow. If you take a look at the world of science today and compare it with science in the 1950s, you'll see that the entire profession has slowly been becoming more "humanized" and less of the tortured, self-denying bunch of objectivists of the sort that Ayn Rand would have advocated. Scientists are more openly human today, and this fellow was a keen observer of the changes early on.

His name was Sir Peter Brian (or P. B.) Medawar. He won the Nobel for his work in physiology revealing the role of the immune system in tissue transplants. He died in 1987 after being awarded virtually every major honor possible in the world of science. He was, of course, a prolific writer of scientific works, but he also had a great many other interests in life, including opera, cricket, the philosophy of science, and the role of science in society in general.

As part of that last interest, he wrote a very interesting short article in 1963 titled "Is the Scientific Paper a Fraud?" He answered his question with a resounding *yes*: "The scientific paper is a fraud in the sense that it does give a totally misleading narrative of the processes of thought that go into the making of scientific discoveries."

Specifically, he took issue with the way the standard scientific paper is written—and, guess what, remember how I told you how clearly structured a screenplay is for a movie? Well, it's the same deal for a scientific paper—the same three acts. Most scientific papers are written according to a *very* strict template that consists of four sections: introduction, methods, results, discussion. But those sections are the same as the three-act structure.

Act One—the introduction, in which the current state of knowledge is laid out, at the end of which, ideally, the knowledge is brought together into a specific question that needs to be investigated (the equivalent to the "inciting incident" in screenplays) and a hypothesis is proposed.

Act Two—"things happen" as methods are described and then the results of the experiment are reported in a completely impersonal way. "Just the facts, Ma'am" is the basic tenet of the second act.

Act Three—the more human element is brought in as the facts collected (the graphs and tables of data) are analyzed and the so-called hypothetico-deductive method is applied to make sense of what just happened and synthesize it into the grand scheme.

In the same way that the movie's lead character pulls it all together at the end of the story—whether it's Rambo laying out his philosophy of life or Bill Murray as John Winger in *Stripes* summarizing his patriotic sentiments—both the third act of a movie and the discussion section of a scientific paper are the place for the grand synthesis. It's basic storytelling dynamics.

What Medawar complained about was the charade the science world has engaged in over the ages in pretending that science is conducted with robotic processes that are not contaminated by the irrationalities of human thought and bias. The inherent philosophy of a scientific paper is the assumption that science is conducted through the process of deduction—that the scientist blindly goes about gathering information on all aspects of a subject and then eventually sits down and, through the process of deduction, puts the information together to create a picture of what's going on.

The truth is, scientists are humans, and from the outset they rely on "induction," which draws on the highly human faculty of . . . intuition. What this means is that a scientist doesn't sit down and come up with twenty-five hypotheses to explain, perhaps, why only the top branches of a given tree species bear fruit. The scientist realizes from the outset that most of the possible hypotheses simply are not likely (the hypothesis that trees learn this behavior from their parents—not an idea really worth testing, you know) and quickly narrows them down to just a few reasonable ideas.

And this is where Medawar says the scientific paper is a fraud. The scientist tells himself that he is writing up "just the facts," but there have been a huge number of biases from the beginning of the project's conception. Which is why Medawar suggests that, instead of saving all the subjective elements for the discussion at the end of the scientific paper, the scientist should open the paper with them:

> The discussion which in the traditional scientific paper goes last
> should surely come at the beginning. The scientific facts and sci-
> entific acts should follow the discussion, and scientists should not
> be ashamed to admit, as many of them are apparently ashamed
> to admit, that hypotheses appear in their minds along uncharted

by-ways of thought; that they are imaginative and inspirational in character; that they are indeed adventures of the mind. What, after all, is the good of scientists reproaching others for their neglect of, or indifference to, the scientific style of thinking they set such great store by, if their own writings show that they themselves have no clear understanding of it?

It's an interesting dilemma scientists face. They have to strive for objectivity and removal of bias—science loses its meaning if it turns into nothing but unsubstantiated opinions. Yet scientists, and the public in general, have to be reminded over and over again that science is conducted by human beings, not machines.

This was what made Stephen Jay Gould's writings and speeches so great. He was constantly pointing out that you can't understand the work of previous scientists without understanding the person and the time in which they lived. All science is conducted by mere mortals.

Scientific Writing Is Still for Robots

Now let me twist this theme a bit with my personal realization that there are benefits to scientists' robotic tendencies.

In late 2006, after I had conducted a dozen or so successful public screenings of *Flock of Dodos*, I wrote up a cutesy two-page list of suggestions on how to stage a screening. I filled it with my usual corny attempts to be fun and folksy and sent it to Steven Miller, my scientist friend who was the executive producer of the movie. I expected a simple "Looks great" reply from him. Instead, what came back the next day was my document, covered in the red corrective ink available in the editing mode of Microsoft Word. He had slashed, burned, rewritten, and restructured the entire essay, removing all my personal tidbits of folksiness, rewording it in a more cold, clinical, professional voice—the voice of the science world.

My blood pressure skyrocketed as I looked over what he had done, and then I erupted with a bitter and angry "How dare you!" e-mail to him. But as I calmed down over the next few days, something began to sink in. Back when I was a scientist—particularly when I was a graduate student—we learned to write first drafts of our scientific papers, give them to colleagues, and then eagerly await their comments. The more red ink on the manuscript when it came back, the better. The only thing that would ever cause anger would be someone *not* covering the manuscript in red ink, suggesting they were just lazy. You never wanted to hear a short "It's great" reply, other than from your parents.

And so, when I got to film school, as we began writing scripts my first instinct was to ask my classmates to read my scripts and comment on them. But I noticed something immediately: almost no one else did this. The other students were terrified to have other people read their material. And, guess what, after a while, I developed the same fear. What was going on?

Well, my big realization is that one of the great benefits of having everyone write like robots in the world of science is that you end up being able to evaluate each other's work like robots as well. Because the scientific paper has so little of the human element (it's created only from the head—no heart, no gut, and *definitely* no sex), when you read it you don't feel very much. And when you write your comments on it, you don't feel very much. And when the writer reads your comments, that person doesn't feel very much, either, except gratitude for your taking the time, or disappointment if you don't have much to say.

In harsh and severe contrast, creative writing is exactly the opposite. For creative writing to be good—for it to reach inside people as it's supposed to do—the writer has to infuse every sentence with the energy, vitality, and life of the writer's personality. Creative writing draws on all four of the organs, in a big way. As a result, the entire process is massively personal. It has to be. People tell you, "Don't take it personally," but that doesn't work for art. If you're creating true art, you have to take it *all* personally because it's your personality you're seeding the work with.

This explains my horrible reaction to Steven Miller's comments on my quirky little essay on conducting screenings. He, still being a scientist, thought nothing of giving it a major rewrite. And, had it been fifteen years earlier and the essay a draft of a scientific paper, I would have thanked him sincerely. But instead, it was about art. And I reacted like a diva.

"Reality Ends Here"

Now let's go back to the power of storytelling. The USC School of Cinematic Arts, when I was there, was made up of two main buildings with a walkway connecting them. Scratched into the cement of that walkway, from long ago, was the motto of the film school—"Reality Ends Here."

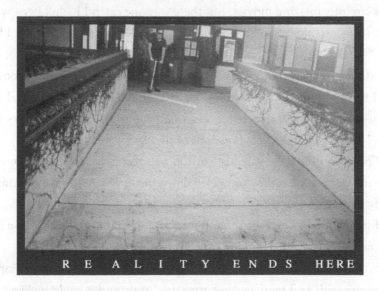

Figure 3-3. The motto of the USC School of Cinematic Arts, scratched into the walkway between the main buildings. Photo by E. Schmotkin.

I saw it the first day I arrived at the school, in January 1994, and had a light chuckle. It would take me a few months to begin to grasp how powerfully true the little slogan was.

I brought to film school what I thought was the creative equivalent of gold. I had fifteen years' worth of amazing stories from the world of marine biology: typhoons, shark attacks, sinking ships, modern-day pirates—a treasure trove of stories. I had spent a month in Antarctica, not just diving under twelve feet of ice into the clearest water on Earth but also spending night after night in the dining hall listening to military helicopter pilots tell about crashing into icebergs, rescuing survivors of ice-crushed ships, and hovering over enormous pods of humpback whales. And I had spent a week in an undersea habitat, living at a sixty-foot depth and listening to the tales of the support divers, who had worked on oil rigs, battling sharks and watching drowned divers' skin turn to foam as they rapidly ascended from extreme depths. I had heard about divers dragged down by weights into black, watery graves.

My mind was overflowing with these stories, and I went to film school to turn them into amazing movies. But there was one catch I hadn't been warned about. It was scratched in that walkway—"Reality ends here," or, in blunter terms, "We don't give a sh*t about what really happened."

It didn't take long for me to end up in a writers' group, telling about the novel I had written that was set in Antarctica, cobbled out of all the real things I had seen and heard about there. And when I finished telling the story to the group, I looked up and saw expressions of disappointment.

"Could you have the scientists at the one research base have a bunch of automatic weapons that they smuggled down there, which they use to attack the scientists at the neighboring base?" someone asked.

I listened for a moment and thought about what a stupid suggestion it was. "Well, no," I replied, "that would never happen."

"Why not? Why couldn't they mutiny at their base and go on a rampage?" the one guy said. And then another guy said, "Yeah, and it could be like *Lord of the Flies*—they're all isolated and have lost their minds."

Another guy chimed in, "And the genetic work they're doing has suddenly given them superpowers." On and on, as I sat there thinking, "My stories aren't good enough. Reality loses to fantasy when it comes to telling stories." And that's the bottom line. The harsh truth of it. Yes, the supposed reality of reality television shows is hugely entertaining, but those things are concocted. The actual, real reality of the real world—the ugly, non-archplotted details of the tedious day-to-day life of scientific research—really doesn't cut it for the broader audience.

Why did they want to add these elements? Because "it makes for a better story." Slowly but surely in these writers' groups and classes, I began to realize there is a partnership at work. Audiences have a set number of stories that they like to hear and that they want storytellers to tell. If you can lock onto one of these set stories, all of a sudden everyone can really start to have fun.

That's what I was talking about with my crafting of the story for *Flock of Dodos*. When it was just my story—my telling of things the way they looked to me—the size of the audience was very limited. But when I began to rework the details into *their* story—into something that resembled one of those standard stories *they* like to hear (about the hero defending the damsel from the neighboring dragon)—I began to enter into the agreed-upon partnership that exists for storytelling.

It truly does work this way. One interesting guy I got to know in my early Hollywood adventures was Danny Sugerman, former manager of the 1960s rock band the Doors. He and Jerry Hopkins coauthored *No One Here Gets Out Alive*, the tremendous biography of Jim Morrison, lead singer of the Doors. They consciously set about taking the events of Morrison's life and making him into a myth—a character larger than life. They did a wonderful job of it, the book became a best seller, and the legend of Jim lives on. But Danny did have to contend with one big annoyance: the Doors' guitarist, Robby Krieger, regularly gave interviews in which he said that Morrison wasn't that big of a deal; he was just a human. Nobody wants to hear that.

Similarly, doctor-turned-science-fiction-novelist Michael Crichton describes in his autobiography, *Travels*, how when he was first becoming famous while still a resident at Harvard Medical School, his colleagues would ask breathlessly what it was like in Hollywood. If he told the truth and said it wasn't that big a deal, they were disappointed. They wanted big, grandiose stories. As he put it, "The blatant insincerity of the way I was treated troubled me very much. I didn't yet understand that people used celebrities as figures of fantasy; they didn't want to know who you really were, any more than kids at Disneyland want Mickey Mouse to pull off his rubber head and reveal that he's just a local teenager. The kids want to see Mickey. And the doctors in the cafeteria wanted to see Young Dr. Hollywood. And that was what they saw." People like their big stories. It's a natural part of being human.

Accuracy versus Boredom: The Two Mistakes of Storytelling

So now we know that scientists can be very, very good at maintaining their sobriety and making sure that, whether or not they enjoy the story being told, they don't get so swept up in the magic that they allow the storyteller to make big mistakes. But is accuracy the only important challenge a storyteller faces?

The answer is, of course, no. A storyteller faces two big challenges: to keep it accurate *and* to keep it interesting. If either is not attended to, errors will result. We'll call these two errors "type one" and "type two," in part because I want to eventually draw a parallel with the world of statistics.

To a scientist, there is nothing worse than inaccuracy. The *American College Dictionary* defines science as "a branch of knowledge or study dealing with a body of facts or truths systematically arranged and showing the operation of general laws." The key part is "facts or truths." Science means nothing if it isn't grounded in the truth. This is why scientific fraud is held up as unforgivable. It seems people can commit plagiarism in other disciplines and get a slap on the wrist, but in science, to be caught fabricating data is to have lost all meaning to the profession and to be banished for the rest of time.

As a result, scientists watch movies about science with an eagle eye for every single detail. And the makers of films about science live in dreadful fear of hearing from scientists that they "didn't get it right." When you hear scientists complaining about Hollywood's portrayal of them, the complaints are always along the lines of "That's inaccurate—that's not what we really do or sound like." What you don't hear much complaint about is the second fundamental error—mistakes of boredom.

What is boredom? It's the state of being bored. What is the state of being bored? It is to experience something that is dull, tedious, repetitious, uninteresting. So it's the opposite of interesting. And to "be interesting" is, according to the closest dictionary I can find . . . "to arouse a feeling of interest."

There's that word again, "arouse." It's about stimulation. Something that is interesting stimulates the neurons in the brain. Something that's boring doesn't. And when the brain is numbed into disinterest, communication doesn't take place.

So what's worse, to communicate inaccurately or not to communicate at all?

It's the dilemma that scientists and science communicators face every single day with every communication exercise they attempt. And that is because accuracy and interest do not always go easily hand in hand.

Let's get back to those two errors that I mentioned in terms of telling a story. They are similar in nature to the two main errors that statisticians worry about in their work.

When scientists need to make a decision ("On the average, is this species of tree bigger than that one?"), they bring in the statisticians. These are the folks who enable us to say confidently, on the basis of numbers, whether the decision is "Yes, this one is bigger" or "No, this one is not bigger."

With any given decision like this, two fundamental errors can be made. The first error possible, known generally as a type one error, refers to the idea of making a "false positive." In a legal case, it would be basically the risk of hanging an innocent man. In the case of scientifically describing nature, it would be the risk of saying you see something when in reality it isn't there.

The second error possible, obviously known as a type two error, refers to the idea of making a "false negative." In a legal case, it would be the risk of letting a guilty criminal go free. In the case of nature, it would be the risk of failing to see something that exists. For example, the two tree species really are different in size, but you don't have enough data to draw that conclusion, and thus you make the mistake of concluding that they are not different in size.

The most important thing to note for these two errors is that we don't live in a perfect world. Which means that it is rarely possible to do a good enough job that you can guarantee not making either mistake. To deal with this, we end up choosing one of them as being more important than the other. In the case of the legal system in the United States, we place highest priority on the type one error. We say that, all else equal, we're more concerned about punishing innocent people than we are about letting guilty people go free. And so we have a default rule that a suspect is innocent until proven guilty. A nation could just as easily have the opposite legal system—that those arrested are assumed guilty until they can prove otherwise. The key point is that you have to choose one. It's like in baseball, where "a tie goes to the runner." It has to go one way or the other.

So here's where it gets interesting for communication. We see the same two types of errors for storytelling (errors of accuracy versus boredom). The choice must be made which of the two errors is most important. Yes, I know, you're thinking, "I want both—a story that's accurate *and* interesting." That's ideal, but in the real world you still have to choose one, just as you do with the two errors in statistics.

And when the dust settles, it's clear that scientists, being detail oriented and believing that accuracy is sacrosanct, will always focus on errors of accuracy as their greatest concern. In the same way a physician lives by the credo of "First, do no harm," a scientist lives by "First, make no mistakes of accuracy." And this is a great strategy for the ivory tower, where the rules for decision making are yes, no, and later. But, as the amount of information and the pace at which it is communicated increase in our society, "later" is becoming less of an option for the science world. And that leads to a major, major, major quandary, which I shall now address through an extremely important case study.

Case Study: Two Global Warming Movies of 2006

Before I begin this discussion, I want to make my overall opinion clear concerning Al Gore's movie *An Inconvenient Truth*. It is, plainly and simply, the most important and best-made piece of environmental media in history. End of story.

You can talk about Rachel Carson's *Silent Spring* and how it gave birth to the entire environmental movement, but Al Gore's movie took the broadest and most urgent environmental issue and jumped it up from background noise to buzzword. There's no point talking about any shortcomings as if they mattered. You can expect only so much from a single piece of media. His movie went way beyond what anyone could have realistically expected. In the spring of 2006, when I was at the Tribeca Film Festival with *Flock of Dodos*, I heard skeptics in the independent film world laughing about Al's movie being "a PowerPoint talk—who's gonna want to buy a ticket to a movie theater to see that?" Most of them couldn't believe it when the movie scored over $50 million in worldwide box office. It was an unmitigated success that deserved to win both an Academy Award and a Nobel Prize, and, guess what, it did. Total success.

However. That said, it doesn't hurt for us to take a few minutes to compare it with another movie on the same subject that came out the same year from the same executive producer.

In April 2006, HBO aired a documentary about global warming titled *Too Hot Not to Handle*. It is a very solid, relatively impersonal and objective effort featuring interviews with a lot of top scientists. It aired on Earth Day and came out on DVD a few months later.

In May 2006, the feature documentary *An Inconvenient Truth* premiered. The movie is a personal narrative by former vice president and Democratic presidential nominee Al Gore about his lifelong connection to the topic of global warming, dating back to his undergraduate days. Interwoven with his PowerPoint presentation of the impending risks of global warming are personal insights, in which Gore reveals the pain of tragedies involving his sister and his son, as well as occasional humorous quips.

Here's where it gets interesting. In addition to their subject matter, these two movies have one large element in common—the executive producer, Laurie David, was a major mastermind of both movies. She's the former wife of comic writer and actor Larry David—co-creator of *Seinfeld* and star of the extremely funny HBO series *Curb Your Enthusiasm*.

Laurie David is herself a force of nature. She has been a board member and trustee of the Natural Resources Defense Council for years and has collected mountains of accolades for her relentless work on global warming. So it is fascinating to compare these two films, not just in terms of substance and style but also in terms of our two potential errors—accuracy and boredom.

The HBO film has tons of substance. It is packed full of scientists talking, and when experts of their caliber talk, they are incredibly accurate. They know their stuff. So it scores an A+ on substance, and you can be certain the accuracy is high. In terms of style, it was well shot and well produced, with plenty of beautiful images of nature to illustrate what the scientists are talking about. But when you consider the question of whether it's boring—well, it doesn't have a personality associated with it. It doesn't tell any sort of intriguing story. It mostly just disgorges the facts and details and lets them splat on the floor for everyone to pick among. Bottom line, it is pretty boring.

The Al Gore movie is sleek, cool, and as hip as the formerly dull vice president could possibly be packaged. It scores pretty close to an A for style. And when it comes to substance, it has plenty. That's why it won an Oscar—it's rich in both substance and style.

Gore didn't shy away from wading into one graph after another, in a manner no one has ever had the courage to do when hoping to reach the general public through film. So it is a reasonably unboring movie (though I'm sure thousands of schoolkids who have been forced to watch it would disagree). But when it comes to accuracy . . . *that* is where it gets interesting.

The Al Gore film is not 100 percent accurate. Countless opponents of global warming science have made as much hay as they possibly could out of this. But both sides agree there are shortcomings in accuracy.

Perhaps the most reliable assessment comes from Danish biologist Kåre Fog on a website comparing the number of "flaws" and "errors" in the Gore movie with those in the books written by Bjørn Lomborg, one of the most prominent in the chorus of voices who are skeptical of environmentalism. It's a fairly balanced assessment that, if anything, is probably skewed in Gore's favor, since the site is so anti-Lomborg. But even Fog's analysis concludes that there are at least two "errors" (things that are factually incorrect) and twelve "flaws" (he defines a flaw as "a misleading statement which does not agree with the facts").

The *New York Times* gave an overview of the science community's assessment of Gore's film. Perhaps the most important opinion in the article is that of James Hansen, director of NASA's Goddard Institute for Space Studies and the leading critic of the George W. Bush administration's handling of global warming. The article says, "Hansen said, 'Al does an exceptionally good job of seeing the forest for the trees,' adding that Mr. Gore often did so 'better than scientists.' Still, Dr. Hansen said, the former vice president's work may hold 'imperfections' and 'technical flaws.'"

When those words come from such a powerful scientific source, who is desperately fighting the fight for global warming concern, you know there genuinely are "errors of accuracy" in the movie.

And yet, when we look at a few simple indicators of the "success" of the two films, what do we see? As I write this, looking at Amazon's DVD sales, the HBO movie is ranked just over 35,000, while the Al Gore movie ranks 431. And when I look at their respective pages on the Internet Movie Database (www.IMDb .com), I see that the HBO movie lists just 2 external reviews, while the Al Gore movie has 357.

Guess which movie had the greater impact? Try asking your neighbors which title they recognize. One of the two reviews listed for the HBO movie actually compared the two movies, side by side, and had this to say, referring to Davis Guggenheim, director of the Al Gore movie:

> While Guggenheim's film is split more evenly between biography and science, and *Too Hot Not To Handle* is more heavily weighted towards the facts and figures, it's not the most compelling presentation. It's the cinematic equivalent of brussels sprouts vs. chocolate: one is good for you and certainly something of which everyone should partake, but the other is definitely tastier and more appealing.

Table 3-1. *Too Hot Not to Handle* versus *An Inconvenient Truth*

	Too Hot Not to Handle	*An Inconvenient Truth*
Amazon rank	35,000	431
IMDb external reviews	2	357

Now, here's the most important detail of all. I spoke with one of the scientists in the HBO movie. This scientist told me that when it came to that movie, Laurie David did a very conscientious job of getting everything right scientifically, at great cost in time, energy, and entertainment value (as the review above indicates).

But when it came time for the Al Gore movie, "she basically asked all the scientists to leave the room," this scientist told me. She simply said that global warming is too important a topic to allow it to get bogged down in facts, details, minutiae, excessive attention to detail, and poor storytelling.

They went for it on the Al Gore movie. They made a film that scored over $50 million in box office worldwide, that was not totally accurate yet is still endorsed by James Hansen and most every other major climate scientist, and, most important, that changed the world. What do you say to that?

TOO HOT NOT TO HANDLE
vs.
An Inconvenient Truth

Figure 3-4. What did *Too Hot Not to Handle* and *An Inconvenient Truth* have in common? They had the same executive producer, Laurie David. *Too Hot Not to Handle* was accurate but not popular. *An Inconvenient Truth* was popular but not accurate. Guess which one was full of scientists.

The *New York Times'* final word on the subject:

> "On balance, [Al Gore] did quite well—a credible and entertaining job on a difficult subject," Dr. Oppenheimer said. "For that, he deserves a lot of credit. If you rake him over the coals, you're going to find people who disagree. But in terms of the big picture, he got it right."

You Choose: Accurate but Not Popular, or Popular but Not Accurate

I will never, ever endorse the idea of striving for anything less than 100 percent accuracy in the making of any film related to real issues in the world of science. My movie *Flock of Dodos* has no scientific "errors" in it. But it also has very little science content, particularly in comparison with something as bold as Al Gore's film.

Nevertheless, this is the fundamental dilemma facing the world of science today. What are you going to do about this movie that turned out to be the most important piece of environmental media in history yet is not completely accurate?

The major scientists agree that the movie's errors are minor and do not change its overall message, which they feel is completely accurate. But still, if you wrote a scientific paper and it was revealed that your data points in a graph were fudged even just a little bit to make your graph appear more convincing, you could say "Bye-bye, tenured professorship."

There's a fundamental disconnect here, and given the idealized, objectivist, rational-thinking values that true scientists cling to, I don't think they ever want to have to deal with this dilemma. But it's there, it's real, and it's as fundamental to the communication of science as the bonds that hold molecules together.

So the big question remains: What are you gonna do when you finally realize there is more than accuracy involved in the effective mass communication of science?

Go ahead, tawlk amongst yerselves on that one.

One More Handicap for Scientist Storytellers

Here's a quick story about coupons. Once upon a time, when I was in high school, my father sat down at the breakfast table, opened a new box of Wheaties cereal, and began pouring the contents in his bowl, but all that came out were paper coupons. He was late for work anyhow, so he got up, disgusted, threw the box on the table, muttered something about "stupid products," and stormed out.

My mother and brother stared at the box for a moment and then closed in. There was no cereal in it. Something had gone wrong on the production line. The only thing it had was about 1,500 coupons. Each one was worth one point, and you had to collect 100 to win a free turkey. Thanks to some freak accident, we had enough for fifteen free turkeys! (I swear to this; you can ask my mother or brother.)

We cashed some of them in at our local store, kept a few, and gave a few to friends and the rest to our church. My father never really quite got it. I don't think he ever made it past his anger about the absence of his cereal. But this tale bears relevance to science communication and storytelling.

My father was as mad at his box of Wheaties as a lot of today's scientists are at the attacks on science. And yet, just as my father stormed away from a potential opportunity, it is the same story when scientists try to shut down the attackers of science. They are missing a valuable communication opportunity—a chance to tell a good story.

The Heart of a Story Is the Source of Tension or Conflict

One of the simplest rules they drilled into our heads in film school is that "the heart of a good story is the source of tension or conflict." Read any standard book on screenwriting: this is what it will tell you. And this is the major source of problems for most boring movies—no significant source of tension or conflict.

Do you think those science "educational" films I got subjected to in junior high school told stories? Of course not. They were just facts, facts, facts. No conflict. Nothing at stake over how it will turn out. It was like watching paint dry. You're not worrying about whether it will dry—you pretty much know it will. There's just no story to hold your interest.

A good story begins at the end of the first act. That is where the tension is established. For the first part of a movie, we usually get to know some sort of place and people. Everyone's happy. Right about the point where you start to think, "Something had better happen or I'm changing the channel," something usually does happen—the monster comes to town, the husband cheats, the child is kidnapped. Basically, the audience members sit up in their seats and say, "Whoa, this looks like a good story."

Here again is where overly literal-minded scientists go wrong. They look at the people attacking evolution or global warming science and they get furious, wanting to shut them down and prevent the public from hearing them. But all you have to do is look at the number of times the subject of evolution has appeared on the cover of *Time* and *Newsweek* magazines in the past five years. It was hardly ever on the cover during the previous decades, but suddenly the conflict brought about by the intelligent design movement turned the subject into a good story, making it of interest to a broader audience.

The attackers of science are a potential communication opportunity. They are a source of tension and conflict. They can actually be used to tell a more interesting story, one that can grab the interest of a much wider audience. Which is exactly why I've used them in both of my movies.

Concision and the Elevator Pitch

So, now that I've overstayed my welcome with this chapter, I will finish with a few words about keeping things short. In the excellent 1992 documentary *Manufacturing Consent: Noam Chomsky and the Media*, the legendary linguist and political activist Noam Chomsky complains about his battles with the medium of television, specifically the news talk shows that won't have him as a guest.

Over the years, Chomsky has learned about a criterion that television producers call "concision": the ability to speak in concise sound bites and not go on and on and on, like I did with Spike Lee once upon a time. Chomsky views it as a conspiracy—television producers end up using that criterion to decide whether they want you as a guest—not whether you're the world's top expert on the topic they are covering but whether you are able to shut up when needed. Chomsky does not accept this idea of keeping things brief, and in the movie he seems proud of and rebellious about it.

Well, what he calls a conspiracy, I call just plain common sense. It is a basic conversational skill to be able to listen while talking so you can recognize when you're boring your audience. A lot of intellectuals, once again preconditioned from too many years of lecturing to prearoused students, have lost this ability to self-edit. Judging from Chomsky's comments in the documentary, he is one of the worst.

And this brings us to the idea of the "elevator pitch": the ability to explain your project, whatever it is, so succinctly that you could get all the way through it in a single elevator ride. How do you do this most effectively? By having a clear structure to your information, using the basic three acts I've talked about.

You set up your subject (first act), give it the twist at the end of the first act (first plot point), explore several possible ways to untwist it and relieve the tension (second act), reveal a possible solution (second plot point), and then weave all the content together to release the source of tension (third act).

Something like this: "I study a starfish on the California coast—the only species that spawns in the dead of winter. I thought it might be due to predators of the eggs being less common at that time of year, then I thought it was due to the best timing for the spring algae bloom, but now it looks like it probably has something to do with a seasonal migration of the starfish, which is what I now study—the way that spawning season might be related to adult movements of starfish."

"Starfish on the California coast" is the first-act setup. "The only species" is the establishment of tension (sets up the question "Why is it different?"). "Predators" and "algae bloom" are the multiple themes of the second act. "Seasonal migration" is the relief of tension, and "what I now study" is the third-act wrap-up.

And there's the shorter version, for the single-floor elevator ride, which is only a single line—"I study the one species of starfish that spawns in the dead of winter instead of during the vibrant spring season." That's enough to establish the sort of tension ("Why is this starfish different?") that will leave the listener still thinking and interested when you step out of the elevator on your floor.

This shorter version is the same as what is called "high concept" in Hollywood, the telling of an entire story in a single sentence or phrase. I'm sure you've heard the ultimate example of this—"snakes on a plane," which actually ended up being the title of a mediocre 2006 movie. It's usually the mixing of two simple elements, each of which tells its own story—"snakes" signifies a dangerous thing that you'd better not let loose; "a plane" signifies a confined space in which you wouldn't want something dangerous loose. The combination instantly fires up your imagination, which is the goal of a good story.

For the elevator pitch, "spawning starfish" signifies something that needs to happen when things are alive and conducive to the survival of the spawn, and "dead of winter" signifies the worst time of the year.

Concision versus "Dumbing Down"

One of the criticisms of *Flock of Dodos* (as I've mentioned, there were many—particularly from the science bloggers) was that I was advocating the "dumbing down" of science. But I was actually trying to do just the opposite.

Let's look at the difference in these two terms. "Dumbing down" refers to the assumption that your audience is too stupid to understand your topic. So you water down all the information or just remove it, producing a vacuous and uninteresting version of what in reality is complex and fascinating. "Concision" is completely different. It means conveying a great deal of information using the fewest possible steps or words or images or whatever the mode of communication is. The former results in a dull, shallow presentation; the latter is a thing of beauty that can project infinite complexity.

Just ask a mathematician about concision. It's the difference between the clumsy mathematician who needs 100 steps to solve an equation and the skilled one who can do it in 5 simple steps. The latter is arrived at either by genius or by hard work. And that's all I'm advocating for science communication—that you be either a genius or a hard worker. That you accept that poor communicators are able to say the same basic things as good communicators—they just need a lot more time and space in which to do it, which ends up boring everyone.

Neil deGrasse Tyson and a Well-Told Story about *Titanic*

So let me tie this chapter up into a neat little package by showing (rather than telling) the real-world power of a well-told story.

Astronomer Neil deGrasse Tyson is the sort of natural born storyteller that the science world desperately needs. I attended a Hollywood event where he spoke and showed what I mean. The event was put on by the National Academy of Sciences as part of its new Science and Entertainment Exchange program, which is an effort to help improve the accuracy of science in movies and television.

Tyson talked about the movie *Titanic*. At the end, when the ship has sunk and everyone is floating in the North Atlantic Ocean, you can see the stars in the night sky above them. But when he first saw the movie in a theater, Tyson noticed something very troubling. As he explains it, there are only two sets of stars the moviemakers could have put up in the sky—the right ones (the Northern Hemisphere constellations) or the wrong ones (the Southern Hemisphere constellations)—so they had a fifty-fifty shot of getting it right. Guess which one they chose.

He said it spoiled the movie for him (typical scientist!), but a couple years later he was walking down the street in New York City and happened to randomly spot the director of the movie, James Cameron. He introduced himself and politely told him of the mistake. He said that Cameron took it in for a second, thought it through, and then sarcastically said, "Gee, I bet if we hadn't made that mistake the movie would have made a couple hundred million more at the box office."

But there's more to the story. Tyson said that in 2005 he got a call. It was one of Cameron's producers, who said they were re-editing the movie for the ten-year anniversary DVD edition, and "Mr. Cameron said you have some suggestions for us about our stars."

Now, that is a good story. Three months later, I told it at the beginning of a workshop on storytelling. Two days after that, at the end of the workshop, without forewarning and after subjecting the students to a two-day "information storm" of lectures and discussions on a wide variety of subjects, I closed the workshop by asking if anyone could remember anything at all about Tyson's "*Titanic* story." What they said surprised even me. A woman recounted the story with complete precision. And most everyone else in the class, while impressed with her performance, said they could probably have done just about as good of a job. Bottom line: it made them believers in the power of a well-told story.

Tyson's story is so effective in part because it has the basic elements of three-act structure. It has a beginning that sets up the theme (inaccuracy of science in big-budget movies), it has a middle that takes us to the opposite place from where we were hoping to go to (the hopelessness of it all when Cameron ridicules the inaccuracy), and an ending that is truly uplifting and satisfying (the sign of hope for humanity when it turns out Cameron was in fact troubled by the inaccuracy).

That is the power of storytelling laid bare. If you can encapsulate your message to the general public in a story as amusing, as compelling (with clearly dramatic highs and lows), and as concise as that, you could . . . well, for starters you could maybe end up as popular and effective a science communicator as Neil deGrasse Tyson!

Being able to tell a concise, interesting, and entertaining story that also conveys substance is a trait that everybody likes. And that brings us to our next chapter, on the importance of creating a likeable voice.

Therefore: Update for the Second Edition

ABT to the Rescue

As I said in the previous chapter update, over the past eight years I've gained a clearer understanding of the fundamental divide between the non-narrative and narrative worlds.

This chapter is now kind of fun to look back at and see how I was swinging in the dark, not yet knowing about the ABT—the tool I would need to understand this divide. I had written about the development of this tool in the introduction to this book and presented it in great detail in my 2015 book *Houston, We Have a Narrative: Why Science Needs Story*.

Two places where you can see this are in the two simple story lines I give at the end—I didn't know as I wrote them, but both are ABTs. The three-act story about starfish research I present starts with a few facts (the AND elements); says, "BUT now it looks like . . ."; and then states the consequence/THEREFORE part, "which is what I now study. . . ." Same thing with the Neil deGrasse Tyson story about *Titanic*, which is actually two ABTs. See if you can spot them.

The most important part of this chapter is my "zero-to-hero" account of editing my documentary feature film *Flock of Dodos*. The early drafts were painfully boring, but we eventually turned it into something that aired on Showtime to a mass audience.

Now, eight years later, armed with the ABT, I can explain what happened much more concisely. We basically went from the boring AAA (And, And, And) non-narrative structure to the engaging ABT (And, But, Therefore) narrative structure. The former brought misery; the latter brought joy.

That, in three sentences, is pretty much the essence of what I took six pages to explain previously. Among many attributes, the ABT provides this sort of shorthand for discussing basic narrative structure quickly and simply.

These days I'm practically obsessed with the ABT. It's the core of the Story Circles Narrative Training program I now run, which is spreading (as of this writing we've involved more than 1,000 scientists). In the near future I will write an entire book on the ABT, as there's now so much to be said about it.

In fact, recently Jayde Lovell (my Story Circles co-creator) hit on perhaps the most enormous validation of the ABT to date. We were at the end of a call and, trying to be funny, she said, "Okay, call me maybe?," referencing Carly Rae Jepsen's 2012 mega-hit song. Suddenly she went quiet as she quickly sent out a tweet. Try singing the song to yourself—you'll realize what she tweeted. The main chorus says, "Hey I just met you AND this is crazy, BUT here's my number, SO call me maybe?" "So" is a word of consequence equal to "therefore." The music video has over a billion views on YouTube. The ABT is at work everywhere in the world.

So I say the ABT is a miracle AND has infinite power, BUT I'm getting old and fuzzy-headed; THEREFORE I invite you to challenge me on it. If I were twenty-five years old right now, I would set to work creating an entire body of literature that would someday result in my being honored as the founding father of ABT analysis. But there's a problem preventing this: I'm not that young anymore and, more important, I really, really love surfing—a lot more than battling with academics.

Shifting the Presidential Baseline

I predict there will eventually be a moment of victory for the ABT similar to what we received this year for the term "shifting baselines." As I explained in the previous chapter, Jeremy Jackson and I launched a campaign in 2002 to promote wider understanding of the term throughout society. It started with my op-ed in the *Los Angeles Times* that November, which eventually ended up on scores of blogs and in at least three college textbooks.

At the start of our campaign, we said our ultimate goal was "to someday hear the president of the United States say the term 'shifting baselines.'" In the spring of 2017 we achieved that goal—kind of.

The president of the United States, talking about the vanishing ospreys on the Potomac River, said, "That's what scientists call shifting baselines syndrome—each generation accepts their generation's version of nature, plunders it, then leaves the next generation to accept the depleted version, and so on." It was a perfect, broad summary of this very important conservation principle.

So now you're thinking, "Wow, that's the smartest thing I've ever heard Donald Trump say." Well, um . . . okay, time for the truth. Sorry—those words were actually spoken by President Conrad Dalton, played by Keith Carradine on the CBS show *Madam Secretary*. Obviously, one of the show's writers had taken a good ecology class in which they'd talked about shifting baselines. I guess, given that Dalton was only a fake president, this is "fake news" in terms of achieving our goal, but so what. At least 9 million viewers got to hear that bit and say to themselves, "Um . . . shifting whuts?"

A Final Word from the *South Park* Folks

As I mentioned in detail in *Houston, We Have a Narrative*, I was first inspired to create the ABT when I heard the co-creators of the animated series *South Park* describe the Rule of Replacing (see chapter 5 for more about this). Shortly before that book came out, I mentioned it to my longtime Hollywood mentor Mike Backes. He's the guy who guided me into the crazy acting teacher's class, and he's the all-knowing being who gave me powerful guidance in the editing of *Flock of Dodos*.

Mike mentioned that he's buddies with Matt Stone (one of *South Park*'s two brilliant co-creators), and he offered to e-mail him the chapter describing how I adapted the ABT from the Rule of Replacing. A couple weeks later Matt wrote back a great e-mail and said it was fine to share with audiences this bit, which has become a mainstay of my talks. Referring to the ABT, Matt said:

> Hey man this is really cool. If there is anything Trey and I work hard on it's structure. It's so important and so so hard to get right. We beat ourselves up so that causality is really there or at least emotionally implied. Then we put in some dick jokes and poop stuff and, voila!

At first glance it seems like a bit of their standard silliness. But when I looked more closely at the message, I realized that it's a perfect reflection of the core philosophy of our Story Circles Narrative Training. Notice that he says structure is "so" important, but "so, so" hard to get right. That is the core of our Story Circles approach.

Communicating effectively doesn't happen through one-day workshops tucked in here and there (and it definitely doesn't happen by listening to the obfuscation of the National Academy of Sciences). It requires a serious, long-term commitment. Just like sports. Just like Brian Palermo and The Groundlings with improv. Story Circles involves one hour a week for ten weeks.

Developing narrative intuition takes time—something that is in precious short supply these days in our overly rushed society.

CHAPTER 4

Don't Be So Unlikeable

One night in Hollywood, a few years after film school, I hopped in the car with two former classmates, Jason Ensler and Jay Lowi, and Jay's girlfriend, Courtney Ashley. It was Friday night, we were headed to a party in the Hollywood hills, and all of us were feeling optimistic at the end of a good week. Jason was being considered as director of a television show, Jay was up to direct a movie, and I had a feature script that several companies were thinking about optioning. With a big smile, I said from the front passenger's seat, "You know, by Monday afternoon we could all be major players in Hollywood!" Before anyone could agree, Courtney spoke up from the back seat. "Now, listen," she said hesitantly. "I don't mean to be ... how shall I say this ... 'the Randy' of this group, but ... I think the odds of that happening are awfully slim." The car was silent for a moment as we processed this. Then everyone burst out laughing, and continued laughing for the rest of the night, as a new expression was born that we still use to this day. Whenever someone is deluding himself with excessive optimism and losing touch with the cold, scientific facts of the real world, somebody invariably chimes in, "Now, listen, I don't mean to be 'the Randy' of this group, but ..."

Like a Completely Dreamless Sleep

It's time to add up everything we've been through so far and consider how it affects the most important variable—whether the people in the audience like what they're hearing. You can avoid the pitfalls of the first three chapters (be not so cerebral, be not so literal-minded, and tell a good story), but if there's something about you that people don't like, you're still not going to communicate very effectively. So let's begin by examining the unique aspects of the scientist's persona.

A fun piece of knowledge for marine biologists is that the great American novelist John Steinbeck spent much of his life probing the interface between science and society. Though he was a man of letters, his very best friend in life was a scientist—Ed "Doc" Ricketts, a marine biologist. Ricketts not only fascinated Steinbeck but also is believed to have been the inspiration for, and at times the source of, much of the author's deepest and most contemplative writings. Some biographers of Steinbeck have even hinted that, at one point, when he was most fed up with the viciousness of literary critics and the superficiality of the world of fame that engulfed him, Steinbeck toyed with the notion of chucking it all and going off to study sea creatures full-time with his marine biologist buddy. When Ed Ricketts died in 1948 in a car-train collision, Steinbeck was jolted and said, "He was part of my brain for eighteen years."

In 1940, at the height of their friendship, but in what some believe was Steinbeck's winter of his own discontent, he took a sea journey with Ricketts. Steinbeck's second marriage was unraveling, Ricketts was trying to escape a failing relationship of his own, the planet was on the brink of world war, and the two men sought a spiritual escape through the ocean. The book Steinbeck wrote upon his return (much of it cribbed from Ricketts' journal), *The Log from the Sea of Cortez*, is considered by many to contain some of Steinbeck's most complex and sophisticated writing.

In one of my favorite passages, so relevant to the persona of a scientist, Steinbeck recounts the story of Jimmy Costello, a Monterey news reporter who was called to the beach one day on the news that a "sea serpent" had washed ashore. A wave of excitement had swept the town. Jimmy managed to find the creature and the assembled mob on the beach.

He rushed, approached the evil-smelling monster from which the flesh was dropping. There was a note pinned to its head which said, "Don't worry about it, it's a basking shark. [Signed] Dr. Ralph Bolin of the Hopkins Marine Station." No doubt that Dr. Bolin acted kindly, for he loves true things; but the kindness was a blow to the people of Monterey. They so wanted it to be a sea serpent. Even we hoped it would be. When sometime a true sea-serpent, complete and undecayed is found or caught, a shout of triumph will go through the world. "There, you see," men will say, "I knew they were there all the time. I just had a feeling they were there." Men really need sea monsters in their personal seas and the Old Man of the Sea is one of these.

Steinbeck goes on to describe the mythical Old Man of the Sea and then comes back around:

For the ocean, deep and black in the depths, is like the low dark levels of our minds in which the dream symbols incubate and sometimes rise up to sight like the Old Man of the Sea. And even if the symbol vision be horrible, it is there, and it is ours. An ocean without its unnamed monsters would be like a completely dreamless sleep.

Let's take a moment to savor the beauty of what Steinbeck meant. It is the downside of scientific thinking—the evisceration of all that is mystical, alluring, mesmerizing, and elusive about life. It is the snapping on of the light in the dark room where the imagination is running wild, only to shed the light of science and reveal that no, there really isn't anything too crazy going on here. It's just a basking shark. No need to indulge your fantasies.

Had my friend Courtney been there, she would have told Ralph Bolin to word his message as "Now look, I don't mean to be the Randy of this group, but there are no sea serpents and this thing is just a basking shark that we've all seen a thousand times before."

And the whole crowd of townspeople would have taken this in, looked at one another for a moment, and shouted back at Bolin, "*Oh, don't be such a scientist!*" Then they would have gone back to calling it a sea serpent.

This, once again, is representative of the standard psyche and public presentation of the scientist. It is a negating role: the designated driver, remaining sober amid the fantasy-blinded townspeople. Truth teller is a valiant role, and I am in no way saying scientists shouldn't play it. When the spread of sexually transmitted diseases can be prevented by the use of condoms, we certainly don't want scientists to hesitate in advocating them simply to avoid spoiling the romance of sex.

But in the end . . . it's just not a very likeable trait. Nobody likes a party pooper. So the question is whether there are ways to play the scientist role without being the negating, annoying, no-fun voice. Essentially, is it possible to be the scientist and still be liked? That is the focus of this most important chapter.

Rising Above

Before I get to the more positive side of things and address the subject of being likeable, let me begin with the inverse—how to be unlikeable, and what the consequences are.

It's time to go back to that acting class that so thoroughly spun my head around. Yet another basic principle was ground into us night after night: all else being equal, audiences do not like characters who "rise above." To rise above is to condescend, talk down to, be arrogant, act superior. All of those things. They are unlikeable traits. The audience no likey.

To see this up close and personal, we did exercises in which we acted out two different approaches to the same problem. The first involved rising above; the second didn't.

So again, we're back in acting class, this time with two students playing a husband and wife at home. The husband has just discovered that his wife stole $100 from him.

In version one, he is furious, can hardly contain his anger, treats her in a degrading manner, and calls her all sorts of terrible things. He says, "You stupid, lying, thieving tramp. You're a scumbag. Give me back my money or I'll slap you!"

It doesn't take long for the audience to start feeling sorry for the woman, figuring there must have been some reason for her to take the money and that he should try to be more understanding. Before you know it, it's the criminal (the wife) who has the sympathy of the audience.

In version two, the husband avoids rising above at all costs and tries to come down to the wife's level or lower. He speaks softly to her in a pleading, compassionate, understanding voice. "How could you do this? We've worked so hard to build a trusting relationship. I just don't understand. You must have a good reason for it, since I know you would never do this without one. Still, this hurts me very deeply."

In this version the audience feels for him, turns to her, and practically wants to speak for him, saying, "Yeah, you rat, how could you do this?" And she is left with a great temptation to say, "Ah, quit your sniveling," which is, alas, rising above, and the end of her with the audience.

It took me a while to really take this in and digest it. But when I did, I realized the value of this dynamic in the real world. I can tell you that, since watching those exercises, I have put it to use several times in actual conflicts.

The fact is, when something has gone wrong, and especially when people know they are guilty of something, they are defensive and ready to escalate immediately. All you have to say is "You lying thief" and the other person will shoot up higher with "Well, who are you to talk—remember last year when I caught you . . ." and the whole conflict spins out of control.

But this is where you do have a certain amount of control over your own destiny. If you manage to restrain yourself, you can take the confrontation in a completely different direction. And I have done it—actually surprising people who expected me to erupt in anger.

Just last week, a friend caught a student cheating on an exam and asked for my advice. I said, "Instead of rising above and scolding this student, see what happens if you come down to his level—make it clear you were hurt that he did it and ask why he did." Sure enough, the student ended up pouring his heart out, crying, and agreeing to withdraw from the class on his own. The professor had been braced for all-out warfare and was shocked by the different outcome.

In fact, my entire movie *Flock of Dodos* is one big exercise in *not* rising

above. I interviewed people whose ideas and viewpoints I consider largely fool-
ish and illogical yet restrained myself from belittling them. Many viewers have
asked, "How did you keep from shouting at them?" and critics complimented
the film for avoiding the attacks one would expect from an evolutionist.

The acting classes paid off.

Rising above is a guaranteed path to conflict, which is sometimes okay, but
it's important to know that it is not the only way to approach an issue—and,
more important, people don't like it.

Just look at your typical villain. What's the most common trait, for everyone from Hitler to Dr. Evil? It's the arrogance of believing they are smarter and
better than the rest of the world. It's a repulsive trait—a guaranteed pathway to
not being liked. It doesn't matter how smart you are. And, in fact, it's character-
istic of some smart people.

British filmmaker Adam Curtis created an excellent documentary on the
history of public relations titled *The Century of the Self*, which focuses in its first
half on Sigmund Freud's nephew Edward Bernays. The man was a genius at
mass communication, applying many of Freud's ideas to marketing and founding the profession of public relations in the 1920s. But, according to his daugh-
ter, he could also be arrogant, condescending, and hateful. She recalled, "It can
be a little hard on the people around you. Especially when you make other people feel stupid. The people who worked for him were stupid. And children were
stupid. And if people did things in a way that he wouldn't have done them, they
were stupid. It was a word he used over and over and over—dope and stupid."

I'm afraid that is the same kind of language I heard coming out of the
highly educated evolutionists in the summer of 2005, and it helped prompt
me to make *Flock of Dodos*. They called the proponents of intelligent design all
sorts of names. I thought to myself, "Don't they realize how they're coming off?
Are they clueless about such simple interpersonal dynamics?"

Unlikeability on Display: A Debating Faux Pas

Here's another example of perceived arrogance, whether intended or not. A
while back, a public debate over global warming was held in New York City. Two

teams of three experts squared off over whether global warming is a crisis or not. Before the debate, the organizers polled the audience about which side they agreed with; then they repeated the poll when it was over. During the debate, the two sides argued vigorously, and there really didn't seem to be an obvious winner in terms of content presented.

However, when it came to style, there was one defining moment, according to the three audience members I spoke with—a moment that swung the tide for anyone who was undecided on the issue.

It was in the middle of the two sides arguing a point. The moderator zeroed in on one of the members of the "global warming is a crisis" team, asking him why he felt his opponents were misrepresenting the issues in their presentation. The scientist replied, "I don't think they are completely doing this on a level playing field that the people here will understand." The audience didn't like this.

It wasn't a malicious thing to say. There were no bad intentions, and I hate even citing this incident, since the scientist is a really good guy. I could just as easily have said the same thing without thinking. But, in a venue like that, per- ception is reality. And the comment was perceived to be condescending. It was a very educated Manhattan audience, and they didn't take kindly to it. On the recording of the event you can actually hear people objecting in response.

When the debate ended, the second poll revealed a shift of about sixteen percentage points against the "global warming *is* a crisis" team. And while it's true that the organizers had to some extent "stacked" the audience toward one perspective, there is no denying how this incident played. According to all three of my sources, you could feel the mood shift right at that moment. They felt that the vote had little to do with the substance of what was presented that night. It mostly came down to style and an unfortunate moment of talking down to the audience.

Style becomes so much more powerful than substance in large public venues with broad audiences. And this brings us to a fundamentally difficult dynamic.

Science Is, Unfortunately, a Negating Profession

I touched on this a bit earlier, in chapter 3. Now I'll go into it in depth.

The entire profession of science has at its core a single word, and that word is "no." Science is a process not of affirming ideas but of attempting to falsify ideas in the search for truth. This is what a hypothesis is—an idea that can be tested and possibly falsified and rejected.

When you give a scientist a paper, he or she reads it with the assumption that the writer is guilty of being wrong until proven innocent. The writer proves his or her innocence by either presenting data or citing sources. With each statement made in the paper, the scientist reading it says, "I'm not sure I believe this." As the author presents graphs and tables of data and cites sources, the good critical scientist attempts to falsify what is being said.

Eventually, after the scientist has examined the data, looked up the cited sources, and found that in fact, despite considerable effort, the hypothesis presented cannot be falsified—only then does the scientist finally start to relax a bit and say, "Well, okay, I think I can probably live with this."

Tough business. It really is. As I waded through my first decade of rejection in Hollywood as a filmmaker, people would ask me whether I found the rejection hurtful or depressing. And I would respond, "Are you sh*tting me? Do you have any idea what it's like to deal with the rejection of scientists? Hollywood folks reject things on the basis of the idea that 'it just didn't grab me,' and they can't even articulate the reason for their decision. When scientists reject you, they hit you with a stack of data and sources that are the basis for it. That's the sort of specific, substantive rejection that truly hurts."

A Critical Thing Called "Critical Thinking"

This negating approach gives rise to something known as critical thinking, which I believe can, to a large extent, be learned. I say this because of a wonderfully stupid experience I had as an undergraduate.

I grew up as a God-loving (sort of), Kansas-raised, young Republican imitation farm boy (I was raised in the suburbs of Kansas City but knew some farm boys) who transferred from the University of Kansas to the University of Washington in Seattle halfway through my undergraduate career. There I fell in with a bunch of sandal-wearing, establishment-questioning, nonconforming (though looking very much alike) hippie graduate students in biology.

One night, at a dinner party blanketed by incense, candles, and patchouli stench, I got into an argument about politics that quickly turned into me, the frat boy from Kansas, versus them, the unwashed masses. Somebody finally interrupted me to say, "How do you know that's true?" And I very smugly and confidently shot back, "I know it's true because I read it in *Reader's Digest.*"

Well . . . you can imagine what happened. There wasn't the slightest trace of a smile or humor on *my* face—I *was not* joking—but the entire dinner party erupted into a screaming, howling pack of jackals, rolling on the floor, peeing in their pants, slapping each other on the back, as I sat there angrily shouting, "What? What? What's so funny?"

Bastards.

It proved to be a moment of awakening. A voice in the back of my head asked at that moment, "Is it possible that not everything written in *Reader's Digest* is correct?" I'd honestly never considered that.

Don't worry; I went to graduate school at Harvard and got cured. And the world of science was the biggest part of my "awakening." Which convinces me you can actually learn to be a more critical thinker ('cause I sure was a dumb Kansas bumpkin back then).

So in the beginning, when I was a junior at the University of Washington, I would read a paper for a discussion group, think, "Wow, what a great and exciting study," and then attend the formal discussion with the graduate students, every one of whom, as they filed into the room, would say, "This week's paper is a piece of crap!" And my smile and enthusiasm would immediately wither into nods of agreement, saying, "Yeah, it really sucked," even though I had no clue why.

That was when I first began to learn the idea of reading a scientific paper with "no" being the starting point.

Fifteen years later, as a professor at the University of New Hampshire, I would bring a half dozen graduate students into my office to discuss recent research papers and realize the tables had turned. The students would come in saying, "That was an interesting paper we read this week," and I would scowl at them with a look of "What do you mean?" Then I would tear into the paper, pulling it apart at the seams, showing them how poorly designed the hypotheses, experiments, and analyses were, and accusing the writer of presenting a discussion that was totally bogus. And on some days they would cower and glare at me as if to say, "Dude, don't be *such* a scientist."

It's a problem. It's at the core of the entire world of science. And it can, and frequently does, run amok. You meet scientists who have lost control of this negating approach to the world and seem to sit and stew in their overly critical, festering juices of negativity, which can reduce down into a thick, gooey paste of cynicism.

You should have seen some of the department meetings I sat through in the zoology department at UNH. Good ideas would be presented and then ground to shreds by the fifteen or so professors present who proudly "poked holes in it." Finally, whoever had presented the good idea would leave with the feeling that it wasn't such a good idea after all.

Crucified by a Scientist

I got my own dose in graduate school of how badly scientists can lose control of their critical approach. A professor on my thesis committee was notorious for his destructive behavior with students. And he did it with me. He turned what was to have been a routine one-hour oral exam—where professors ask you a few softball questions just to let you show off how much you know, and then pass you—into a five-hour ordeal of humiliation and frustration. The worst experience of my academic career.

The same professor demolished the careers of numerous graduate students. One close friend quit her doctoral program after five years of work because he did the same thing—only worse—to her in a committee meeting. As she was showing slides and presenting her research to her committee of five professors, he flipped on the lights and said, "I want you to stop right there and tell us what distinguishes you, in the way you're doing this research, from a common hired laboratory technician."

What do you say to that?

She quit. He went on. He mellowed over the years and eventually became less destructive. But there are countless stories like that; I've listened to so many over the years. In the last discussion I had with Stephen Jay Gould, he commented on what a shame it is that it takes so much positive energy to build up a student's inspiration and drive, yet only a single negative experience to obliterate it all. It's not an even balance, as I'll discuss in a minute. But first, a few words about this spectrum that exists in scientific thinking.

Science, like art and most other professions, requires a mixture of two elements—creativity and discipline. Science without creativity is dull, but science without discipline is dangerous. And here we are again, back to these two key elements—the objective and subjective parts. Discipline is the rigid, regimented, more robotic objective component that has to be brought to bear for science to work properly. Wild ideas are fine, but without discipline they become a waste of time and energy. Creativity is the more human, liberated, unrestrained element that must be let loose for it to work. Science without at least a little bit of creativity is just plodding detail that does not expand our understanding of the world.

But at the ends of the spectrum—at the far ends—lies darkness.

Creativity, unleashed with no restraints and allowed to shoot too far out on that end of the spectrum, eventually results in sloppiness. This is the classic mad scientist stereotype. You see it in the real world of science. You can usually spot it in the scientist's office—you walk in and there's junk piled everywhere, hundreds of cartoons and meeting badges and photos plastered all over the walls with no organization whatsoever. And you ask for a copy of a paper by the scientist and he spends the next fifteen minutes exploring stacks of papers while talking to himself and discovering manuscripts that have been lost for weeks. Just like Doc Emmett Brown (Christopher Lloyd) in *Back to the Future*.

That's the funny part of sloppiness. But the more tragic part is when those scientists give talks at scientific meetings or write papers and their data are as big a mess as their office. Then it's not so funny. It becomes sad, depressing, and dysfunctional. Particularly when you watch the better scientists rip into them in public during the question session at the end of their talk.

At the other end of the spectrum lies an even more destructive excess. Discipline shows itself in critical thinking and the ability to organize the scientific process effectively. It is essential, but just on the other side of discipline lies this abyss, this quagmire, this Hades known as cynicism.

Table 4-1. Creativity and Discipline Spectrum

I------------------------------I	------------------------------------I	------------------------------I
Sloppiness Creativity	Discipline	Cynicism

In its lighter forms it's funny and even somewhat healthy. But in its most concentrated state it becomes a toxic miasma where not even the existence of a soul can be seen. It's a dragon, content only when it has managed to breathe its fire of negativity across the rest of the sparkling universe. And that force of negativity is the handicap that dogs the world of science when it comes to mass communication.

The Core Problem: No Doesn't Equal Yes

So what's so bad about negating? Being known as a tough critical thinker sounds like a good thing. And when you watch a group of top scientists get together and critically analyze a proposed idea, doing what they are best trained to do, it can be an impressive spectacle—like a group of competing alpha males pounding their chests and proclaiming dominance as they grind up what previously sounded interesting.

But it's a different story when you take that behavior out from behind closed doors. What is admired within the cloisters of academia can be horrifying when unleashed on the general public. And that's because the masses thrive not on negativity and negation but on positivity and affirmation.

Don't believe it? Just watch *The Oprah Winfrey Show*. What do you see, day after day? Stories of hope and joy, uplifting, inspirational, fulfilling . . . the kind of stuff that makes scientists want to vomit. But there you have it. Like night and day. Scientists versus the rest of the world.

Just look at the most popular movies. They're mostly inspiring stories of hope. Not a lot of blockbusters that end with the hero plowing his truck into a school bus full of kids.

Now you're thinking, "But science is fun! I've even seen that slogan on buttons from the National Science Foundation. Kids love science. It can be uplifting, too."

Well, yes, in the right hands and presented the right way. But let it not be forgotten that deep in the belly of the beast that is science resides a ferociously destructive force. And what is scary is that the force is both powerful and unifying. Let me tell you about it.

The True Believer

I'm no long-term fan of conservative writer P. J. O'Rourke, but he did nail one of the most insightful essays on the American environmental movement with a piece he wrote for *Rolling Stone* for Earth Day 1990. He gave it the brilliant title "The Greenhouse Affect," and he drove his cynical SUV-like voice right into the heart of the environmental movement, pulled out a bullhorn, and captured the core of the worst side of environmentalism.

What he talked about is the incredible unifying force created by negativity and hatred. Citing Eric Hoffer in *The True Believer*, he said, "Hatred is the most accessible and comprehensive of all unifying agents. Mass movements can rise and spread without belief in a God, but never without belief in a devil." He then recounts the famous anecdote of a Japanese diplomat sent to Berlin in 1932 to study the National Socialist movement. When asked what he thought of Germany's social agenda, he replied, "It is magnificent. I wish we could have something like it in Japan, only we can't, because we haven't got any Jews."

The world of science does have a devil, and that is inaccuracy. Things that are factually wrong are so motivating that I have seen scientists at meetings who I know dislike each other team up to attack a speaker whose ideas they both believe are wrong. It's the basic "my enemy's enemy is my friend" dynamic inexorably come to life. That's how powerful this force of negation can be. But again, when you bring it out into mainstream society, the dynamics just aren't the same. Scientists have to learn to hide their inner Frankenstein when in public. Which they normally do. But, thanks to modern technology, glimpses of this cynicism are now no farther away than the screen of your laptop.

The Joy and Bane of the Blogosphere

As I mentioned earlier, an increasing number of scientists have created their own blogs and are now building a public audience for their raw, unedited, uncensored thoughts of the day. And that's great. Blogs are immensely compelling. They capture that unscripted spontaneity, that this-could-go-anywhere energy, that's so important in engaging an audience. Science blogs can attract people who wouldn't naturally sit down and read *The Origin of Species* but who are intrigued by the frank debates about evolution.

Unfortunately, these debates can get more than spirited—they can get really mean-spirited. Much has been written about the overall tone of bloggers in general—not just scientists. Stand-up comic Patton Oswalt described the problem in rather crude terms on the Comedy Central show *Lewis Black's Root of All Evil*:

> Bloggers have taken one of the most essential human activities—communicating—and degraded it to nothing more than electronic poo flinging. No, no, I'm sorry, that was an insult to the fine poo-flinging community, because at least poo flingers take into account wind velocity, aim, and poo density.

If you take an acting class, you'll get a little insight into the dynamics of negativity. In learning to act (which is essentially a form of communication), anger is the starter/entry/first-step emotion that most actors are drawn to.

Our instructor explained to us from the outset that the real goal of acting is to "lose yourself" in the performance—to become so totally absorbed in the character that you basically *are* that character while you're onstage. The part of your brain that keeps saying "I'm acting, I'm up here acting, this is still just an exercise in pretending" finally gets shut down.

We didn't see it happening at all in the beginning. People would do exercises and you could tell they were just play-acting. But a few weeks into the program, people began to have breakthroughs where suddenly something very different and magical would happen. They would dive so deeply into a performance that, when the instructor finally shouted "Scene!" and walked onstage, the actor would look dazed for a few seconds as he or she let go of the fantasy world and returned to reality. The instructor always picked up on those moments and excitedly shouted to the class, "You see, you see—look at this actor! She doesn't even know who she is right now—she fell so deeply into this character she has to shake herself to come back down to earth."

But here's the interesting part. The easiest and most accessible means of finding your way into that world of real acting was through "negative spirits"— basically anger, rage, hatred, bile, screaming, shouting—you name it.

And, to be honest, the simplest piece of dialogue for breaking through was two words, "*F*** you!*" Night after night we would have to sit and listen to our classmates screaming this over and over again.

We did all those exercises for a year, and then we were ready for the more advanced stuff, including "good spirits"—the exact opposite. That's where we were able to see how harsh the contrast was.

It turns out it's a whole lot harder to convince people you're happy than it is to convince them you're mad. The standard exercise was to receive the news that you'd just won a million bucks in the lottery. You'd be amazed to see how difficult it is to make that moment believable. It's just so much more complex, elusive—hard to even put your finger on it. I only know we sat there night after night watching classmates get the good news and dance around in circles, jump up and kiss their partner, and say, "Yay! Yay! Yay!"—and none of it was the least bit believable.

So here's my take on the blogosphere: the joy and bane of blogs is that the vast majority of their authors are not veteran writers. They are mostly newcomers with little experience in communicating to an audience. And yet they seek to create a compelling and believable voice—one that many people will want to listen to. They face the same challenge as the starting actor.

It's not surprising, then, that they would be drawn to the same entry point in reaching their audience—the immediacy of anger. The new blogger, I would bet, almost always makes his or her first breakthrough—meaning a post that is widely circulated and talked about—not by extolling the joys of daily life but by cranking up the rage and anger, producing the sort of typical "rant" that draws readers in and gives them the feeling that they are listening to a voice that is speaking the truth (and notice that a rant is also a good source of tension or conflict, making it into a good story for people to relate to).

Bottom line—the blogosphere is filled with introductory students, just like in the acting class—all seeking a voice and trying to write something compelling. Anger is their entry portal.

When I realized this, I began to have much greater respect for the bloggers who have moved beyond the elementary emotions of negativity to the higher plane of good spirits (yet avoiding cloying, syrupy "Kumbaya" blather). They are a rare but highly cherished group. When you combine positivity with the spontaneity of blogs, you get a pretty powerful method of communication.

Positivity and Natural Selection

One very simple notion that Stephen Jay Gould drilled into our heads, in the years I spent hanging on his every word, was that of the basic elements of natural selection. He would say it's a relatively simple two-step process. The first step is the creative, nondirectional phase when offspring are created; the second step is the deterministic, much more directed phase in which the environment selects for those with the highest fitness.

This simple conception for natural selection parallels all the two-step/bifurcating/duality phenomena I've described. And it's especially similar to the creative process and the scientific method in general.

When you have a problem to solve, you undergo the same process. First you brainstorm in a totally uncritical, nondirectional way in order to think of all the possible ways you can solve the problem. The more uninhibited you can be, the better the chance of coming up with something brilliant.

Then you enter the second stage, in which you select the ideas that are reasonable. It's basically positivity followed by negativity. And it works well, provided you keep the two separate.

But the positivity stage—the creative stage—is the most vulnerable part. And that is where negativity can wreak havoc. All you need is one person in a brainstorming session who starts saying, "Oh, that's a stupid idea," to immediately constrain the whole creative process. This is when lateral thinking gets inhibited. This is what produces the "typological thinking" that bedevils so many taxonomists—the idea that all species within a group have already been described, and therefore any variation from the established types is simply abnormal rather than a valid description of a new species.

It's a sort of closed-mindedness. And that's what negativity can lead to. Of course, excess positivity can lead to flakiness, but that's way over in the other direction.

All the truly creative scientists I've ever known have had a wonderful aura of positivity around them. They are creative, they are able to discover large ideas, it's a very positive process, and in the end, if you really think it through, you'll realize . . . it's not that different from improv comedy. Just a whole lot of "Yes, and"-ing at work.

The Interview Dilemma: Yes versus No

We now have these two giant forces—the positivity of spontaneity versus the negativity of critical thinking—that seem to be in opposition to each other yet are both crucial to communication. This may leave scientists in a quandary over how to present themselves to the public, particularly when being interviewed.

Should they be the designated driver and make sure not to commit mistakes? Or should they be the fun, lively, go-with-the-flow "Yes, and" good sport who is the life of the party? The former runs the risk of coming off as an unlikeable grouch. The latter can be an absolute disaster, allowing the interviewer to tell the world that global warming will cure cancer and make everyone's lives wonderful.

The solution is, once again, partitioning. Just as natural selection has two phases, just as creativity has two phases, just as science has two phases, so can the interview subject. The bobbing and weaving between modes can create texture and complexity in the interview, instead of it being one-dimensional. A question is asked, the answer begins with a spark of spontaneity—a set of possible answers—and then discipline is imposed.

Q: What do you think caused the dinosaurs to go extinct?

A: Lots of ideas—could have been an asteroid, could have been climate change, could have been too much television, maybe they just got bored, I don't know, but there is a lot of evidence to suggest that one hypothesis is the most logical, which is . . .

There are ways to modulate the answers.

So Are You Telling Us to Get a Hollywood Makeover?

If being unlikeable is a bad thing, then the obvious question is "How can we be more likeable?" I, of course, spent time in acting classes listening to the instructors talk about what makes characters more likeable, and everyone from the beginning of grade school thinks about what makes someone more popular than others. So should scientists just get a makeover and tell everyone what they want to hear?

This question comes up at many of the *Flock of Dodos* screenings, and I see it discussed on blog discussions of the movie. A number of scientists have implied that I'm suggesting that all scientists spend time in Hollywood, take acting classes, and buy new clothes. That's not what I'm saying at all, but it's entertaining to listen to these conservative voices in the science world. They vocalize the ever-present forces opposing change. And it's fun to see them be wrong. Just take, for example, the way scientific presentations have changed over the years.

The Blue Slide Pioneers

I attended my first scientific talk in 1976. Back then, all scientists made their own 35 mm slides for their presentations by drawing graphs on white paper, placing them on a photo stand with lights and a camera mount, and photographing them on standard 35 mm film. The result was projectable slides that looked just like the artwork—black lines and text on a clean white background. But then someone discovered a new way to do it.

It was called diazo processing, and it produced white lines and text on a soothing blue background. Scientists began showing up at meetings and giving talks using these slides, which looked so different. And I distinctly, distinctly, distinctly remember hearing the first responses to them.

Specifically, Jim Porter, a coral reef ecologist from the University of Georgia who was a very flamboyant and lively speaker, was one of the first to use them in my field of marine ecology. I remember standing in the lobby during the East Coast Benthic Ecology Meetings at the University of Maryland in the spring of 1978 and listening to a group of scientists quietly cursing the man—calling him a showman and a huckster and asking, "Who does he think he is?"

It might as well have been a group of church elders fretting over a youngster who was combing his hair the wrong way. Same mass conservative behavior. Everyone subconsciously or consciously attempting to make sure nothing ever changes, regardless of whether it's an improvement on the past or not.

What's funniest for me is that—I guarantee you—*all* of those scientists assembled that day, and at many other meetings where I heard the scoffs of skepticism, are today using the standard PowerPoint slides, which have that same soothing blue background with white letters.

Things do change, but scientists are for the most part programmed, all else equal, to resist changes. And it's particularly difficult for them when it comes to the changing dynamic in communication between substance and style.

Today, Style Is the Substance

It's now time to delve into the core conundrum faced by the world of science. There are many books in the field of communication theory that address this, and I don't have the space or interest to discuss them all here, so I have only one book to point you toward. But it's a good one.

Richard Lanham, in his 2006 book *The Economics of Attention: Style and Substance in the Age of Information*, provides a basic catchphrase—that today "the substance is style." Those are words to live by. What he means is that for every given message there are the same two parts we have discussed—the objective and the subjective, the substance and the style—only he refers to looking *at* the message rather than looking *through* the message. He is pointing to the difference between getting caught up in the style of what is communicated (the "at") and being able to get past style and into the substance (the "through").

Scientists do a good job of looking through the message, into the heart of the substance that's being communicated. But most people never get past the "at" part.

The amazing thing about scientists is that, if you went to a scientific meeting and a speaker got up dressed like a clown, the scientists in the audience, along with everyone else, would initially focus on looking "at" the appearance of the clown. But if the clown began talking about important discoveries and suddenly reexamined a hypothesis in a new and important way, the scientists would be able to get beyond looking "at" the clown and would actually look "through" the clown's appearance to hear the substance of what was being said. And if it were true and correct, the scientists would have no problem having a serious discussion with the clown afterward. Seriously. I swear this is true. All you have to do is go to a scientific meeting and see some of the weird people (and I say this lovingly). They might as well be dressed as clowns.

There is one very famous scientist who for decades has dressed like a homeless person, maintained an unkempt beard, and picked his nose unashamedly *during* his talks to hundreds of scientists. Over the years his audiences have just gotten used to it. The man is brilliant. There's no more reason to focus on his nose picking than to get hung up on Stephen Hawking's computer-generated voice. Scientists know how to do this. It's sort of the upside of being literal-minded—the ability to focus only on what matters.

In contrast, going back to the clown, members of the general public would never get past looking "at" his appearance. Later they would talk about what a fool he was. End of story. Didn't matter what was said. Never got past the big red nose and floppy shoes.

That's the difference between substance and style. In today's world people's minds are cluttered with excessive information. As Lanham describes it, we live in an "attention economy," in which the resource in shortest supply is people's attention. Given such a circumstance, your message means nothing if it isn't noticed. And that just takes us back to my earlier discussion about "arouse and fulfill." It's the same thing—the need to arouse is the same as the need to gain attention. Without it, you're spinning your wheels.

This is where unlikeability can sometimes come into play. It's entirely possible to use it as a tool to gain a certain (albeit limited) amount of attention. You can stand up and insult a room of people, and if you're quick to turn it into a joke, you can actually do a good job of arousing them to what you have to say. But there's a big difference between using a tiny bit of unlikeability intentionally as a tool and just plain being unlikeable.

All else equal, unlikeable traits are simply things to identify and then avoid.

And now I shall launch into the most important element of what I have to say, which provides the synthesis of the previous sections.

That element is . . .

Likeability

The tide goes out; the tide comes back in. The tide has been going out for a long time in this book. It's time now for it to reverse direction.

I've told you about all the ways scientists, and academics in general, can go wrong in trying to connect with the broader audience. And there are surely plenty more ways I haven't even begun to touch upon. But there comes a time to end the critique and answer the question "So what do you suggest we do?"

Let's start with the idea of being likeable. And I don't mean telling people what they want to hear (though sometimes that can be a good idea) or being a sycophant. Those pathways are too direct.

Likeable Electoral Candidates

So who are you going to vote for in the next round of political elections? Will you read the speeches of the various candidates, figure out exactly where they stand on the issues you believe are most important, and assess their leadership skills? Or will you just vote for the one you like best? Maybe vote for the man or woman you saw in television ads who sounded like a pretty decent and level-headed politician. The one you . . . liked.

This is one of the themes presented in *Freakonomics*, the hugely popular book by Steven Levitt and Stephen Dubner. They talk about the widely held belief that, because of the huge amounts of money spent on political campaigns these days, some people believe it's as simple as whoever raises the most money wins. But Levitt and Dubner put this assumption to the test.

They looked at a large number of cases where two candidates have run against each other twice. Just as with William Jennings Bryan, who lost the U.S. presidential election to William McKinley in 1896 and then ran against him in 1900 and lost again, there are many cases where the loser again runs against the same incumbent.

They examined only the instances in which the loser spent more money the second time around, and they found that it made no difference. Money wasn't the factor. Likeability was. When the public doesn't like someone, no amount of money is going to get that candidate elected.

Bubba and Teflon

And, of course, what is the ultimate example of the importance of likeability? For the past three decades, it was Presidents Ronald Reagan (the Great Communicator and the Teflon President) and Bill Clinton (Bubba). Both of them set the standard for likeability in America, and both had an ability to evade the facts of a situation simply by using their charm and charisma.

We'd like to think the general public is so interested, so analytical, and so savvy that they can devour all the long-winded speeches and arguments they are presented on given issues. But they're not. And by "they" I mean myself included. We're all too swamped these days to be able to read and analyze everything.

When the information reaches excessive levels, as Richard Lanham reviews in detail in *The Economics of Attention*, there comes a shift from substance to style—the only way to deal with the "information firehose," as he calls it.

Likeable Lawyers

Of course, this phenomenon isn't limited to politics but can be seen in every area of life, including the courtroom.

My elder brother is a lawyer in Montana, in charge of training the state's public defenders. I sent him an article from the January 2006 issue of *Esquire* magazine that he still uses in his workshops. It's titled "The Drowsy Dozen," by Chuck Klosterman, who argues that the time has come to get rid of the silly, idealistic, antiquated notion of "a jury of peers" and replace it with a system of professional jurors.

He says this because he's sat on too many juries in which two weeks of detailed testimony is presented on forensic science, fiber analysis, DNA testing, and all sorts of other sophisticated science that sails right over the heads of the jurors. At the end of the two weeks the jurors sit in the jury room, totally lost to the science, and end up making their decision on the basis of which of the two lawyers they found most convincing, trustworthy, authoritative, and, basically . . . likeable. Bottom line, they opt for style over substance.

Snap Judgments

Lacking the time and energy to evaluate the information being presented, people end up evaluating the presenter. They are no longer able to transcend style to get to substance. As Lanham says, style becomes the substance.

By the way, the decision of whether you like or trust someone happens very, very quickly. In *Blink*, Malcolm Gladwell cites Nalini Ambady's work on "thin slicing," which focuses on the accuracy of judgments people make based on only short clips of video. She and her colleague James Rosenthal gathered ten-second clips of professors teaching and had students watch and evaluate the professors on a series of standard variables—is the professor warm? enthusiastic? interested? well organized?—as if they were filling out student evaluations for an entire semester. When they compared the results with evaluations made by students who actually did have the instructor for the semester, the correlation was 0.76, meaning that a judgment made in a few seconds was about the same as one made over several months.

Scientists would like to think that people base their opinions about others on a thorough job of getting to know the "reality" of who they are. The harsh truth is that opinions are mostly based on the quick "perception" of who they seem to be. And this means that simple, superficial elements—such as picking your nose or wearing clownlike clothes—can be extremely important.

Yes, that is today's world. I'm sorry to be the bearer of such news, but like-ability is a very important factor. And let me tell you about my own experience with likeability in the television world.

The Power of Jack Black

I mentioned earlier that I wrote and directed a television commercial (or public service announcement, PSA) with twenty comic actors, including Jack Black. He played the lead role as the conductor of our bad symphony for the oceans. When I finished editing the PSA, I hired a distributor, who packaged it, sent it off to 1,000 television stations, and then provided us with "tracking statistics" showing where and when it aired across the country.

The PSA proved enormously popular, airing on roughly 350 stations for a total of well over 30,000 showings, generating more than $10 million in free airtime (television stations choose a select number of PSAs and broadcast them for free).

Figure 4-1. Comic actor Jack Black turned in a splendidly silly performance as conductor of the *Ocean Symphony* public service announcement in 2003. His likeability factor was as big as his eyes. Photo by E. Schmotkin.

Our distributor randomly chose a dozen stations that had aired the spot and then contacted them and ran them through a simple questionnaire. The main question was "Why, given the huge number of PSAs that you receive and the small number you can air, did you choose this one?"

For nine of the twelve respondents, the answer was not that they chose it because of the importance of its message, or the relevance of its message to their viewers, or the effectiveness of its message. No, those are all criteria that would have involved the substance of what we produced.

To the contrary. Nine of the twelve respondents said, "We aired it because we liked it."

It was that simple. They popped the tape into their VCR (it was still pre-DVD days), they watched it, they all laughed, they all agreed they are big fans of Jack Black, overall they liked the spot, and so they aired it.

Now compare that result with the PSA produced by the Less Than One campaign, mentioned in chapter 1. That PSA had a dark, gloomy tracking shot of a bulldozer in a landfill pushing a mountain of garbage. As it faded to black, you heard gurgling of bubbles underwater and the narrator saying, "That's what we're doing to our oceans."

I spoke with a PSA programmer at a television station in Los Angeles. He said they received that PSA, they watched it, they said to themselves, "Our viewers don't want to see that sort of dark and gloomy message on their TVs," and they threw it in the trash. Literally.

The likeable PSA scored over $10 million in free airtime. The unlikeable PSA scored virtually no free airtime.

Am I proud that I had to stoop to making a silly bad symphony for the oceans with comic actors in order to get my message on television? Not really. It's a shame that our society has changed from the 1950s, when the NBC show *Watch Mr. Wizard* played to millions of kids and inspired more than 50,000 Mr. Wizard Science Clubs. But things have changed. We live in a new media environment, with different rules. And those rules make the conveying of substance harder than ever. But not impossible, if you're willing to learn the basic constraints of the system.

So What Makes Someone Likeable?

If likeability came down to a formula, scientists would figure it out and be the most popular people in the world. Of course, it's far too subjective for that. But we do know likeability is inextricably tied to elements arising from those lower organs—humor, emotion, passion.

And you can't overlook the overall role of fun. Edward Castronova made this point in his 2007 book *Exodus to the Virtual World: How Online Fun Is Changing Reality*. In a speech, he said, "Fun is a societal element that governments have yet to fully appreciate." If you can create an atmosphere of fun, there are no limits on popularity.

In the end, it is these human qualities that can reach beyond "the choir" of those who are interested in science no matter what. They can be incredibly powerful in mass communication. And, even as the new media environment has, in many ways, made communicating substance harder, it has also opened up new opportunities.

In the old days, scientists were forced to keep their heads low, their noses to the grindstone, doing their humble research and quietly awaiting the day when a journalist would knock on the door of their laboratory and ask them to explain their scientific work to the world. But that day is now over. New technology has brought about many changes in communications, and in the world of science this could prove to be one of the most profound developments.

With the advent of such innovations as blogs, video technology, and YouTube, a new day has arrived for scientists. No longer do they have to sit quietly awaiting that visitor from the media world. But before they speak out, scientists first must learn to listen. This is the subject of the next chapter.

THEREFORE: UPDATE FOR THE SECOND EDITION

A couple of major things have changed since I wrote this chapter in 2008. The first has to do with the world of communication. Back then, blogs were still the main place on the Internet for angry people to spew rage and hatred. Since then, the wonderful world of science and technology has brought us a faster and more efficient means for this called Twitter. Thanks, science!

Second, as I've said, I have developed a much deeper appreciation for the importance of narrative. This means I would add narrative to the central attributes of being likeable—meaning that being boring or confusing is not likeable; being interesting is.

The Arrogance of Academics Continues

"Rising above." This is what we learned about in acting class and what I warned of in detail in this chapter. I wish I could say it no longer happens with scientists, but that would be a joke.

It happens every day. In fact, I think an example of it is the subtext of what the writer Atul Gawande said to the graduating class of scientists at the California Institute of Technology in 2016, which was presented in the New Yorker. If you ask me, his advice was sadly misguided.

Yes, he told them lots of important things about the dangers of today's anti-science movement, but he also said, "Scientific explanation stands in contrast to the wisdom of divinity and experience and common sense." Nothing to argue with there; it's just a question of context. He was sending a kind of a cryptic signal to the young grads that "mere mortals are subject to the weaknesses of religion and 'common sense'—but with science, you are superior to those human failings."

Which might even be true. But I think it's sending them in the wrong direction. If I were to speak to such a gifted group (not that anyone would ever ask a philistine like me), I would take an opposite tack. I would zero in on that term "common sense" and warn them that they are mostly lacking in it.

That would make for a great graduation speech. I'd be saying, "Not so fast, science nerds—you need to know you're common sense–impaired." And I'd be chased off faster than Ann Coulter at a feminist rally. But it would have come from decades of thinking about the arrogance of scientists—like this wonderful memory from back in my old coral reefs days.

"How Did I Manage to Earn a Ph.D.?"

Had I given the speech, I would have started by flashing back to thirty-five years ago—back to my days at the Lizard Island Research Station, at the north end of Australia's Great Barrier Reef, where I spent perhaps the very best year of my life.

It was months of azure-hued water and lush coral reefs. I lived in a tent with a cinder block floor at the edge of an isolated white sand beach. I would awaken to nothing but the billowing of the tent flaps in the ocean breeze with the sound of distant sea gulls. Such are my memories of my Ph.D. dissertation fieldwork on the Great Barrier Reef.

The director of the research station was an Australian scientist who got into arguments with virtually everyone, including the poor sod who flew up from Sydney to build a walk-in freezer for the laboratory. The director's nickname was "Badger" for good reason. I watched one shouting match in the sweltering tropical heat where the beaten-down worker had droplets on his sunbaked cheeks that were a mix of sweat and tears of frustration. The director constantly referred to him as a "f***wit," right to his face.

A few days later I was in the director's office putting to use my skill of reading letters upside down (my office mate Mark Patterson back at Harvard had showed me how, as a graduate student, you should learn to read letters upside down so that when you're standing in front of your advisor's desk looking down you can read his mail). I read a letter on his desk as he was rattling on about maintaining the station's power generator.

It was a letter to that poor sod. It was one in a series of insulting rants they were trading (yes, before there was e-mail people traded insulting letters where they would wait days for the next round of name-calling). This one opened by saying, "If I were the 'utterly useless and incompetent dickhead' you say I am, then how did I manage to earn a Ph.D.?"

I think that one sentence was my first awakening to this social problem of academics, way back then. He was basically saying, "If I'm so incompetent in your field of expertise (refrigeration technology), then how did I earn a Ph.D. in my field of expertise, marine biology?" Which is like saying, "If I'm such a lousy golfer, then how do you explain my ability to build a tree house?" Total non sequitur. Yet I knew what he was thinking—that highly educated people are "smart."

"A Different Kind of Stupid"

That was 1981. Cut to 2011—a couple years after the first edition of this book came out and I was considering titles for a second book. In all seriousness, the title I began incubating was "A Different Kind of Stupid."

Yeah, I know, it would have really pissed off a lot of people. It would have been truly, 100 percent unlikeable. But I'm still thinking about that title, all the time, when it comes to academics. Maybe I'm saving the title for the National Academy of Sciences.

The fact is, educated people do suffer from a different kind of stupid. I love President Obama and honestly do think he was "the messiah" in his presidency. But he was a different kind of stupid on July 27, 2016, when he said in a speech, "There has never been a man or a woman, not me, not Bill, nobody, more qualified than Hillary Clinton to serve as president of the United States of America."

What Obama said was a stupid statement, of the sort that highly educated people make every day. If the presidency required only information processing, then yes, he would have been right. But unfortunately half of the job calls for people skills and charisma, in which everyone knew she was sorely lacking. Those skills arise not from intellect but from what I have termed the "lower organs."

The importance of the lower organs is the most widely talked-about message of this book. It is my plea for scientists to come down out of their heads and make more use of their lower organs. By this I mean the heart (more emotion), the gut (more humor and intuition), and even the sex organs (even some sex appeal at times).

Business communications guru Nancy Duarte, in her best-selling 2010 book *Resonate*, did a two-page spread on "Randy Olson's Four Organs of Communication," including the Arnold Schwarzenegger photo from chapter 1 of this book's first edition. A reviewer for the UK publication *The Times Higher Education* said, "I'm going to write this review in the style suggested by Randy Olson. This means that I'm going to use my penis." Everyone had lots of fun with it.

Eight years later, the message is still the same—you need to come down out of your head. The cerebral part of science is obviously important. But you've also got to have heart, humor, and intuition (the sex part can still remain optional).

The bottom line is that sitting there spewing out information all day long ultimately doesn't work. It just isn't likeable.

CHAPTER 5

Don't Be Such a Poor Listener

In 1998, cell phones were still fairly new. I was living in an apartment in Los Angeles and bought my first one. I programmed into the speed dial function several phone numbers. One of them was the landline in my apartment, which I labeled simply HOME.

A few weeks later, I had just returned from visiting my parents in Kansas City. I picked up my cell phone to give them a call. They were still living in my childhood home, so I instinctively pressed HOME on my speed dial. As I was waiting for it to ring, my landline started ringing. I quickly picked it up and said, "Hello?"—but there was no one on the line. I said, "HELLO?" again, but still no reply. I finally shouted into it, "I CAN'T HEAR YOU—YOU'RE GONNA HAVE TO CALL BACK!" Then I hung up, irked.

And then realized what I had just done. Portrait of the crazy man, alone in his home, calling himself, listening, at least he thinks, as he's shouting at himself, telling himself to call back, hanging up, annoyed . . . with himself.

I Finally Heard the Crazy Acting Teacher

I am capable of listening. It just takes me twenty years sometimes.

This book opened with the paragraph of profanity-laced bile my crazy acting teacher in Santa Monica, California, spewed at me on the first night of her class in August 1994. The verbal beating I took so thoroughly fried me that, honest to goodness, it took me two decades to finally hear what she was saying. It proved to be the single most important moment of my professional life, but it would have been even more valuable had I been able to "hear" her that night.

What I mean by "hear" is not just being able to regurgitate what is said—like a kid playing a video game who says, "Of course I'm listening—you said I need to do this, this, and this." What I mean is that you actually absorb it and let it change your thought processes.

So why didn't I hear her? Because academics don't listen. Which is actually what she was screaming at me—she said she didn't want to work with "your type" anymore because "you don't listen." How perfect was it that I wasn't listening as she was saying I don't listen?

No, instead of listening, I was busy thinking, "This woman is an idiot. She doesn't know what she's talking about." Which is what lots of academics now think about me as they don't listen to what I have to say about narrative.

How Do You Get People to Listen: Honey or Vinegar?

Over the years, lots of people have hinted to me, as I've offered up my sometimes harsh critique of scientists, the old adage "You'd catch a lot more flies with honey than with vinegar."

But . . . would the crazy acting teacher have ever gotten through to me had she spoken sweetly and softly that evening? What would have happened if she had said, "Hey, kiddo, that's not the way we do things in this acting class. I know you're probably not used to this, but we generally don't ask questions. It's a technique I prefer. You see, I do the talking, and I find that if you listen long enough, most of your questions will get answered. Okay, is that pretty clear?

And by the way, I really want to thank you for taking this class; it's a joy to have you with us!"

What would have happened if she had said that in a soft, comforting voice, instead of what she actually did, which was to scream at me like an f-word-fueled blast furnace?

Even now, twenty-three years later, as I relive that evening, it lingers on in my mind as a thing of . . . beauty. Why? Because I did eventually "hear" her. It just took time to sink in.

Acting Teaches Listening

So how do you learn to listen better? Ironically, it's through the very thing she was teaching—acting. Actors are no good unless they know how to listen. They have to be able to take direction, and they have to be able to pick up cues within a scene. Listening is their life's blood.

The crazy acting teacher no longer teaches introductory acting classes. Shortly after my experience, she moved mostly to administration. The intense Meisner in-your-face style takes a toll.

As she would say night after night, "acting is reacting." Audiences don't want to see you act—running through preprogrammed performances that are identical. They want to see you react—taking in what is said to you by other actors and listening to their words—and showing the audience how they affect you.

Here's my favorite example of how audiences want to see you react. Alec Baldwin on *Inside the Actors Studio* told the story of "the day I realized I'm not a brilliant actor." He was in the play *A Streetcar Named Desire*. In one scene, he finished shouting at Blanche DuBois and stormed over to the screen door to leave, but when he pulled on the doorknob, it came off in his hand.

He stood there, stunned, for a moment, uncertain what to do. In retrospect, he said, he knows exactly what a brilliant actor would have done: punch his fist through the screen door, staying in character, and angrily let himself out using the knob on the other side. The audience would have been enthralled by his reaction, maybe even thinking it was planned.

But, as he said, that's what a genius would have done.

Instead, he simply walked around the edge of the flat on the stage as the audience burst into laughter, and the moment was a complete embarrassment. Instead of "listening" to what had happened in the scene, he simply had it programmed in his brain, "I gotta get offstage," and took the simplest route.

I love that story so much. Audiences want to see you react, taking in what is around you and processing it. And that's why acting classes are such good training for listening.

So here's the painful fact: scientists are terrible at listening. In fact, when biologist Peter Kareiva reviewed this book in *Science* magazine in 2010, he said, "The failure of scientists as communicators is that they do not know how to listen."

As soon as I read that line, I realized that was the one big regret I had about the first edition—I should have written a chapter about listening. Which is why it is such a treat to fix that.

So now I'm going to present three case studies of the failure to listen. They involve a former scientist, a scientist, and a nonscientist.

And then, just about the time you're wishing I'd present something more than just vinegar, I'll give you some honey by talking once again about the ABT narrative template—the tool that can not only strengthen your narrative skills but also lead you to an understanding of the power, importance, and process of listening.

1. Michael Crichton: A Future World for Science

Science fiction writer and former doctor Michael Crichton tried the honey route with the science community. It accomplished nothing.

As I mentioned in the introduction, no human has ever had Crichton's combined knowledge of science and media. He was a potential gift to the science world, but scientists were too literal-minded to hear what he had to say.

His brilliant keynote speech to the AAAS, the American Association for the Advancement of Science, in 1999 was the epitome of this. The presentation was so far ahead of its time. Sadly, it is still the best piece of media coaching for the science community in dealing with today's information-glutted world.

In delivering the speech, Crichton didn't rant, shout, or use profanity. He took the honey route, which probably explains why it was so quickly forgotten. He was his usual dignified, erudite, diplomatic, and gigantic six-foot, ten-inch self.

The heart of his talk had a simple problem-solution form. Not only did he identify four of the key problems the science world has with the media world; more important, he offered up specific, actionable recommendations for addressing the problems.

Here are Crichton's four problems and recommended solutions:

1. Feeling "Used" by the Media

Problem: Scientists complain that reporters are not there to help them but rather simply to use them as a character in telling whatever story they desire, whether accurate or not.

Solution: Instead of fighting and resisting reporters, work with them.

2. Inaccurate Image

Problem: Media outlets are constantly "getting it wrong" on science—both factually and perceptually in the portrayal of science and scientists.

Solution: Set up a service bureau for reporters.

3. Fake News

Problem: Journalists tend to be driven more by the desire to "tell a good story" than to get the scientific facts right.

Solution: Establish a "Good Housekeeping Seal" for reporters "so that your denial has power," as he said.

4. Fake Experts

Problem: A science-based problem arises, but reporters end up quoting so-called experts who have little or no credentials.

Solution: Establish recognizable spokespersons—anointed experts who are respected and used for putting a human face on science.

This was powerful, clear, simple, actionable advice coming from the first person ever to have a simultaneous number one book, movie, and television show. He was painting a picture for a future world in which "scientists understand media."

His future world would have looked like this: scientists would routinely reach out and eagerly work with the media; they would have a network of Science and Media Service Bureaus; there would be a Good Science Seal for reporters; and a team of perhaps ten scientists would be anointed spokespersons for key disciplines.

Let's look at the last element. Perhaps the team of experts could be called the Science Voices. One of them, for example, might be Dr. Kathy Sullivan, the first woman to walk in space and an oceanographer, as the Ocean Voice. She would have a team of experts for various issues in the oceans. She might know nothing about barnacles, but if the barnacles started vanishing from New England coasts, she would know who to consult to get the right opinion and would then voice it to the media through the Media Service Bureaus.

Wet Blanketing to Death

Okay, now I shall revert to the 49 percent of my brain that still thinks like a scientist and explain why Dr. Crichton's ideas were tay-rible in the opinion of scientists. I will speak for all the ankle-biters who pride themselves on unleashing their critical skills to explain why none of this could ever possibly work.

Their criticisms would be simple, obvious—and true. They would voice concerns: Who would pick these Voices? Who would host these Bureaus? Who would blah, blah, blah. And when they were done, they would end up just like so many thesis committee members I watched as a professor, critiquing to shambles a student's work until the student quits science. I wrote about this in chapter 4.

When their critiquing was done, they would proudly dust themselves off and science would be back to the current situation of non-leadership in which nobody on the street can tell you where to turn when a science issue crops up. The result is spokespersons for science such as a politician speaking for climate change science (Al Gore), a television entertainer voicing evolution science (Bill Nye), and self-appointed daytime television health experts promoting health remedies of which more than half have no science behind them (Dr. Oz, Dr. Phil).

Crichton knew what he was talking about. But scientists didn't really know what he was talking about because, as he said, scientists don't understand media.

I'm sure the bulk of scientists who listened to his talk just chuckled and said among themselves, "That guy's tall." He wasn't part of their tribe, didn't speak their dialect, so they didn't listen.

Here's how much of his wisdom was officially acted on: pretty much zero.

2. Dr. Michael Mann: I Have Found the Enemy and It Is Intellectualism

The deafness is deafening. Especially in the academic world.

Academics like to argue. They like to look at statements and try to "poke holes" in what has been said. This is the very thing that caused the crazy acting teacher to go bonkers on me. She was wanting to teach but not wanting to have to deal with my negating brain examining the validity of every single thing she said. You can't teach in the face of that much negation.

The crazy acting teacher told us we were never, ever allowed to ask questions. I immediately—and with great certainty—felt that she was the worst instructor ever: the sort of autocrat who never wants to have her authority questioned.

Two decades later, I finally completely understand where she was coming from. I now seek the same thing from the people I want to teach.

Yes, I know there is the great Socratic tradition of questioning everything and learning through questions. That's fine for intellectual pursuits. But intellectualism is not just the enemy of effective communication; it is the roadblock that diverts traffic out into a field where all the cars drive in circles until they plow into each other and explode.

And that's the real problem. Intellectuals are not only bad at communication; they are even worse at explaining how it works. This is the reason the National Academy of Sciences should just cease and desist on the topic of "Communicating Science Effectively," as I said at the end of chapter 1.

So let me tell you about my frustrations in trying to get one academic to listen. He's been my good friend, kind of, ever since serving on the post-screening panel discussion of my movie *Sizzle: A Global Warming Comedy* in 2009, when I visited his home institution, Pennsylvania State University.

His name is Dr. Michael Mann (not to be confused with the very cool director of the movie *Heat* by the same name, but without the "Dr."). He's a climate scientist and has done a lot of good and important scientific research in his career.

In 1999, he and two coauthors established what came to be known as the "hockey stick" graph representing the rapid rise in planetary temperature in recent years. That simple label for an extremely important data set has been an effective tool for mass communication of the core problem of climate change. But that doesn't mean he's a great communicator.

Mike introduced me to his buddy climate blogger Joe Romm, who is in fact a very skilled communicator and author of a short and elegant book titled *Language Intelligence*. Over the years, the three of us have engaged in numerous feisty group e-mails—usually with the two of them arguing some aspect of communication against me.

In 2016, they threw at me every piece of polling data they could find as they said they were certain Hillary Clinton would be our next president. I fought back with what all my narrative principles were showing: that Donald Trump had a deeper grasp of narrative and understanding of media than any presidential candidate, ever. Joe bet me $100 on the election. I won.

Back in 2010, I did a lengthy interview with Mike on my blog, after which I offered up a few suggestions to him on communication. He plays a prominent role in the climate science community, so he's often interviewed by the media. I tried, but over the years I've seen no evidence he listened.

Since emerging in 1999 as a lead spokesman for climate change science with the hockey stick graph, Mike has also become the lead whipping boy for the climate change skepticism movement. In 2009, when someone hacked the e-mails of a group of climate scientists, it was his phrase "hide the decline" that they held up as their primary target for ridicule and insult.

A couple years later, I attended a taping of HBO's *Real Time with Bill Maher* and met a producer of the show. He said to let him know if I ever had any potential guests to recommend. I set to work recommending Mike repeatedly because of his powerful personal story of being the target of climate change skeptics. In 2015 they finally had him on.

After the show aired, Mike asked me how he'd done. I replied, "Do you want flattery or the truth?" He said he was hoping for a little of each. I chose the latter and pointed out that from the start he'd repeatedly negated Bill's statements. Bill's first line to him about climate change was, "Isn't it settled science?" Mike's reply was, "Well, you don't have to ask me." That's a negation, which is less effective than responding with "Yes." These are basic principles of mass communication.

I pointed out this and a few other items that caused my girlfriend at the time to force me to change the channel a couple of minutes into his interview because she was so bored. Mike went silent in our e-mails. We didn't communicate for a year.

He's not the worst communicator, and it's admirable he's willing to take the time for media work, but he just doesn't listen. Part of my notes in that critique and at least one other (when he invited me to hear him speak in Los Angeles) included detailed instruction on the importance of the ABT (And, But, Therefore) structure in argumentation. When his 2016 book *The Madhouse Effect* came out, I flipped through the first few pages. As I expected, it was all AAA (And, And, And). He never listens.

All of which showed itself in an editorial he published that year in the *Washington Post*. On my blog, I did what I've come to call my "narrative analysis" on the piece using the ABT elements. What I showed was that the essay was little more than twenty statements of fact capped off by an eight-word "therefore" statement, "I would urge these scientists to have courage."

I sent Mike an e-mail pointing out these shortcomings of his editorial. He replied by listing all his colleagues who felt the editorial was powerful and compelling (once again showing zero interest in listening). So I just wrote it all up as a blog post to let the public see a classic example of impotent argumentation from an academic.

And so it goes. In fact, as I write this, I just this morning traded a series of e-mails with him about the two huge hurricanes of the summer of 2017 and the need for clearer messaging from the climate science community about the role of warming waters in the disasters. He does so many interviews that he could make this happen, at least somewhat. I gave him a specific recommendation for messaging. All I got back was the usual "I already know this stuff" reply.

He doesn't know it. He doesn't use it. But . . . maybe I'm wrong, because in 2017 the American Geophysical Union awarded him the Stephen H. Schneider Award for Outstanding Climate Science Communication.

What can I say? Stephen Schneider was about the most boring science communicator I ever had to listen to. Yes, I know he was a very courageous scientist who gave everything to climate science, but he was still boring. I attended a living room event in Santa Monica in 2004 where he emptied the room as he droned on and on about climate science to a short-attention-spanned Hollywood crowd. I remember walking out on the roof patio and joining a group who were shaking their heads at the boredom of what he was saying. How fitting that they named their award after him.

Academics don't listen. I know because I was one of them before I met the crazy acting teacher. I think many feel that listening suggests you don't know something—and that's the last thing academics want to admit.

The Narrative Index

This is the saddest story ever. It doesn't involve the science world, but the consequences of it will definitely impact science. To set this up, I'll first tell you about the Narrative Index—a tool that brought the story into focus for me.

In the summer of 2015, after spending almost four years talking and thinking virtually nonstop about the ABT, I was hit by a very simple revelation. It began with the Rule of Replacing from the *South Park* guys. As I explained in *Houston, We Have a Narrative*, they mentioned in their documentary that every time you replace an "and" with a "but" or "therefore," the storytelling gets better.

It dawned on me that there ought to be a metric as a result of the Rule of Replacing. If you take any given text and subject it to the process of replacing "ands" with "buts" and "therefores" exactly as Matt Stone and Trey Parker recommend, there should be a metric reflecting this. You will have increased the "buts" while decreasing the "ands." The metric is extremely simple—just the ratio of "buts" to "ands."

I call it the Narrative Index, or NI. To calculate it, just copy and paste a chunk of text into your word processor, search for the word "but" in order to find the number of times it appears, search for the word "and" to find the number of occurrences, and then divide the two and multiply by 100 to make it a whole number. It's that simple.

I set to work calculating it for several thousand speeches, essays, debate performances, and books and even some songs. Look at figure 5-1—it shows a few representative values. At one end of the range you see the weekly monologues of comedian and political commentator Bill Maher, who has a team of writers to help him make every essay as aggressive an argument as possible. At the other end we have the obviously non-narrative content of equipment maintenance manuals. They don't exactly need to tell a story or argue anything. Telephone books would score even lower.

Realizing how boring Hillary Clinton's speeches were, I began calculating the NI for all of them I could find, and bingo, there it was. She's an "and-er." Big time. The average of her speeches and debate performances was 14 (the values in the figure are only debate performances, which tend to score higher). That's not disastrous, but great speakers like Lincoln, Kennedy, and Martin Luther King Jr. average around 20.

And then came the shocker—Donald J. Trump, at that point in time, averaged 29 for his combined debate performances and speeches. Some of his scores were well into the thirties.

3. Hillary Clinton: Sadly, Such a Scientist

None of this was much of a surprise when I stepped back and looked at it. Hillary is known as a brilliant policy wonk but not a great communicator. She is basically a scientist—fairly cerebral, fairly literal-minded, not a great storyteller, and . . . painful to say, her levels of unlikeability were scoring just below Trump's as both of them were at record levels.

My fun little discovery about Trump began to shift from novelty to the

Figure 5-1. The Narrative "Teeth" Index. Bill Maher's values come from fifty-two of his monologues at the end of his "New Rules" segments on HBO's *Real Time with Bill Maher*; Donald Trump's, from eleven debate performances; Abraham Lincoln's, from his seven debates with Stephen Douglas in 1858; Hillary Clinton's, from nine debate performances; Douglas', from his seven debates with Lincoln; and climate change skeptic Marc Morano's, from nine television appearances. The leftmost value is the average of four equipment maintenance manuals found on the Internet. (By the way, the six chapters of the first edition of this book all scored above 20, averaging 22.)

source of genuine anxiety. I set to work trying to alert the world about it. Trump, with his simplistic views of politics, ratchets through one ABT after another. He says, "We love the Mexicans, AND we want to be their friends, BUT too many of them are coming in illegally; THEREFORE, we need to build a wall." Or "Our tax code is important, AND we want it to be fair, BUT it needs to be reworked; THEREFORE, I'm going to get the best people and we'll do it."

Hillary, on the other hand, offered up more realistic, complicated, nuanced solutions that would take a long time to get through before she could get to the next ABT. The result was a much lower NI.

I wrote a draft editorial and began trying to get it published, but I hit a brick wall. Every major publication already had its own in-house journalist explaining why Trump was popular. I was encountering the tribal nature of journalism—driven more by social dynamics (especially for editorializing) than by facts. As far as I could tell, nobody took what I was saying seriously. Despite valiant efforts by the publicist for my 2015 book, we had zero success.

By the spring of 2016, I was at the end of my wits. Every few days Hillary and Trump would give speeches, I would run them through the Narrative Index calculations, and the pattern would emerge again. He was drawing on the power of narrative, she was just plain boring, unable to identify a singular narrative for her campaign, and, worst of all, she had poorly crafted slogans.

In March, in an act of desperation I typed into Google, "Hillary Clinton boring." The first article that came up was an interview with James Carville, the brilliant campaign manager and strategist who masterminded Bill Clinton's election in 1992. The title was "James Carville Acknowledges Hillary Clinton Is 'Boring.'" I smiled and said, "Bingo."

I wrote a letter, found a friend with a pathway to get it to Carville, and within a few days was on the telephone listening to him spew profanities about the ineptitude of the Democratic Party. I was listening to the legendary "Ragin' Cajun," just as you hear him on television. He eventually got around to my Narrative Index and the ABT. He said, "I've thought for years there had to be some sort of simple structural underpinning to telling a good story."

He had me send him all my essays and videos. I waited for a reply. Nothing. But a month later I got an e-mail from a guy saying he was part of the Clinton campaign—that they were "fans" of my work (which was diplomatic nonsense) and wanted to get me involved.

Over the next three months I came to realize the guy—as nice as he was— had simply been given the assignment to placate Mr. Carville—basically "call this schmuck and make him feel important." He said that three Fridays in a row he had tried to pitch my ABT notes in their weekly conference calls with all the staff, but nobody took an interest. Why? Because they weren't listening.

It's now well-documented that the staff of Hillary's campaign weren't listening to anyone. A Hollywood friend of mine told me that one of the top directors, whose name you'd recognize, tried to get her to listen to some of his communications advice. She didn't listen.

In December, I finally had lunch with James Carville in New Orleans. I asked him if he ever talked with Bill Clinton these days. He said he did every day, and then five minutes later he stepped outside to take a call from Bill. I asked him what had happened with the election—why hadn't Hillary drawn on all the strategic brilliance he had used in Bill's campaign? With a sad expression, he said, "She wouldn't listen to us."

In fact, here's the saddest thing of all. Carville said that when Bill ran for president he had three main slogans—one of which was "He's With Us." Carville said, "You know what her slogan was—'I'm With Her.' I asked her for one tiny revision—if she would consider changing that slogan to 'She's With Us.'"

It could have shifted the entire detached, aloof tone of her campaign. But she didn't come close. She wasn't listening. It's what heavily educated people do.

What Could You Have Done If You Were Hillary's Speechwriter?

If you were a speechwriter for Hillary, you could have helped the campaign with the ABT narrative template. That's not a delusional statement; it's just a fact.

I sent the Clinton campaign samples of "the ABT at work." Here's one of many sets I sent them—an excerpt of a Hillary speech, followed by an ABT-structured rewrite. The transformation is not stunning; it's just a little bit stronger in narrative structure.

Hilary Clinton Speech Excerpt

I am looking forward to the opportunity to discuss how we knock down the barriers that stand in the way of people getting ahead and staying ahead, starting with the economic ones. My focus is on more good-paying jobs with rising incomes for families and how we prevent corporations from taking jobs out of our country by imposing an exit tax, making them pay back any tax breaks they've gotten. But we also need to be having a positive agenda for manufacturing for small businesses and entrepreneurs.

ABT Rewrite

People work hard in this country and have a right to expect a good life, but they encounter barriers. My goal is to knock down those barriers. How do we do this?

With more good-paying jobs. We also prevent corporations from taking jobs out of our country by imposing an exit tax, making them pay back any tax breaks they've gotten. That's how we start to fix things, but there's much more needed. We need a positive agenda for manufacturing, for small businesses and entrepreneurs.

This sort of structuring over a large amount of text can have an impact—especially if the text is structured not just at this level but also at the larger, over-arching scale. So many other simple things could have been done—like having a clear core narrative.

Trump had a clear, simple narrative theme, with "greatness." Clinton had one, but her campaign staff never managed to identify it and message around it. They could have. But, alas, they weren't listening.

Dinner at a Restaurant with No Sign

But one guy did listen to me—the Ragin' Cajun, who is an adjunct professor of political science at Tulane University in New Orleans. In January 2017 he invited me to speak in his weekly political science class. He's been teaching the class since about 2010. It's legendary. He has fifty undergraduates who are selected out of a large number who apply. He holds the class . . . wait for it . . . in the living room of his stately mansion a couple blocks from the campus.

I had a ball. It was one of the most fun nights of my entire life. He and I sat in two big, comfortable chairs on the landing of the large staircase in the center of the house, with fifty students in folding chairs below us and a screen to one side. They brought me over early as the teaching assistants were getting set up. Carville came in dressed in jogging clothes, drenched in sweat, and wearing a U.S. Marines T-shirt—he had been a Marine. He greeted me and went upstairs.

A half hour later, as the seats were filling up, he came back down and started barking orders at the students like a drill instructor. You could see how much they loved him. The other instructor was a lawyer who runs the course with him. He told me about some of the previous guests, including three former presidents. They've had Bill Clinton twice. The second time, Bill's security detail had to cut him off at the end—he wanted to go all night.

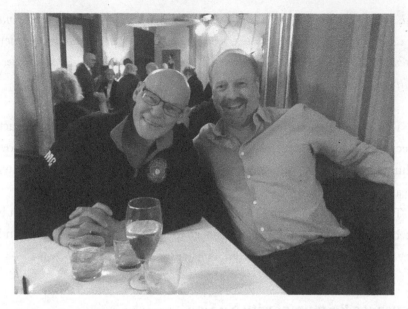

Figure 5-2. It's the narrative, stupid. The Ragin' Cajun, James Carville, exceeds expectations.

Carville asked the students, in his thick Cajun accent (which is the real deal—the guy was born in Carville, Louisiana), if anyone had ever heard of Edward Everett. Of course nobody had, including me. He went on to explain that Everett was the Massachusetts governor who had been the featured speaker preceding Lincoln's Gettysburg Address in November 1863.

Everett spoke for two hours that day; Lincoln spoke for two minutes. "Which speech do you think people remembered?" Carville asked the class, pointing out the core of my message, which is that brevity and simplicity are the soul of pretty much anything effective these days in a world of too much information (did you hear that, National Academy of Science eggheads?).

Then Carville said, "Okay, y'all listen up to what this man has to say," and they did. We spent two hours with me talking and Carville looking stuff up on his laptop (and probably e-mailing Bill and his other buddies) as students asked incredibly bright and challenging questions.

Every five minutes, he would interrupt me to read something from his laptop or pose a question to the students. When I got to my simple definition of the word "narrative" (the series of events that occur in the search for a solution to a problem), he forced me to stop, go back, and repeat it, then again, and then again as he implored the students to "take this down," which they did.

You know what that shows? A true educator who knows the importance of repetition. It's called inculcation.

I delivered my core message, which was that Hillary had no message. Four young women in the second row objected. They had worked on her campaign. One of them said she had quit school the previous semester to live in Brooklyn and work at Clinton campaign headquarters. She strongly disagreed, saying that Hillary's message was always clear . . . to everyone working in the office.

All through the evening, I could hear Carville's wife, Mary Matalin, upstairs talking loudly on the phone. They are famous for being the odd couple of the political world—the Democrat husband married to the Republican wife, both major activists for their parties.

Near the end he went upstairs for five minutes, and then he came down quickly and seemed panicked as he started shouting orders again: "Okay, god dammit, everybody gotta get outta here, we're done, fold up the chairs." It was like the boss had come home and discovered us throwing an illegal party. Watching how fast everyone sprang into action, I could see it was routine for them. I have a feeling Mary has limited tolerance for her husband's rabble-rousing with young Democrats in her house.

Out front in the driveway the students were slowly departing, some still milling around. The four young women who worked for Hillary came up and basically said, "You know what, we got it—we see what you're saying—she really didn't have a clear, overarching narrative." I replied, "Yeah, really sad, but true."

And then came the big treat of the night—two solid hours of storytelling with the Cajun over dinner. We drove about three blocks to another residential neighborhood and parked in front of a nondescript house—he, the lawyer, and I. As we walked to the front door, Carville said, "In New Orleans, if you have to put up a sign in front of your restaurant, it means that it isn't very good."

The door opened and there, almost like some sort of illegal brothel, was a room packed full of elegant restaurant tables and diners. The owner greeted us, every head turned at the unmistakable sight of Carville, and then they seated us at the most central table. What ensued was two hours of incredible stories, many of which were shared with people from other tables who stopped by or leaned over from their tables. It was like a big party with James as the kingpin as they all begged him to find some way to rid the country of Trump.

Suffice it to say, James Carville possesses what I call the ultimate communication goal—deep narrative intuition. He has a mind that is built around the problem-solution dynamic. He has the ability to look at complicated, convoluted content and pull out the simple narrative core.

If you doubt this, watch the iconic documentary *The War Room* from 1993. It's a portrait of Bill Clinton's winning campaign, but the star of the show is James. Over and over again you see people coming to him for advice, not just on strategy but also on messaging, which means narrative. Repeatedly he takes calls in which people with Bill are asking what to do. Off the top of his head he says, "Tell him to say . . ." followed by simple, clear messages, the most famous and eternal of which was "It's the economy, stupid."

James Carville is the embodiment of communication skills. He has narrative intuition, he is deeply impassioned (he cries at the end of that documentary), he has a wicked sense of humor, and he is just plain fun. If only the science world could find a few spokespersons with those traits at the same level.

Maybe someday. In the meantime there is one more, very important lesson to be learned from Hillary's sad campaign. The lesson is articulated by one of the few very cool people I know in the Washington, DC, science establishment. Her name is Shirley.

Shirley's Law

After the whole Hillary Clinton/James Carville experience, I have taken to defining the general public relations problem that occurred as this:

> **Shirley's Law:** If you don't tell your own story, someone else will tell it for you (and you probably won't like what they say).

This simple principle—which most people already know but which needs to be stated formally—comes from the great and wonderful Dr. Shirley Malcom. She is a seventy-one-year-old African American woman who is head of education at the AAAS and fond of introducing me as "my brother from another mother." Actually, what I would call her is my sister from the same academic father, Bob Paine.

Shirley has the most improbable of life stories. She grew up in the deep Deep South in the 1950s and '60s in Birmingham, Alabama, at a time when it was reeling from the struggle for civil rights. As part of the conflict a number of churches were bombed, including her family church, Bethel Baptist. The worst incident occurred after she had left Birmingham for the University of Washington.

On September 15, 1963, white supremacists planted fifteen sticks of dynamite in the 16th Street Baptist Church. The explosion killed four girls aged eleven to fourteen. These girls included the daughters of family friends. Spike Lee's documentary about the bombing, titled 4 Little Girls, contains a clip from the earlier bombing of Bethel, and Shirley can be seen in a photo in the sequence.

As an undergraduate at the University of Washington in Seattle, she studied zoology and was an undergraduate teaching assistant for my greatest science hero, marine ecologist Bob Paine. A decade later he was my undergraduate advisor. He was such a tremendously charismatic scientist that I dedicated the first edition of this book to him. When Shirley spotted that in 2009, we became instant soul mates.

She got her Ph.D. at Pennsylvania State University and then went on to have an accomplished career working tirelessly for women's and minority programs within science organizations and in support of historically black colleges and universities (HBCUs). Today she is the longtime head of education at the world's largest general science organization, the American Association for the Advancement of Science. She was one of the first to "get it" on the ABT narrative template when I explained it to her in early 2012. She gets it on everything to do with communication.

To return to Shirley's Law, this really was the sad case for Hillary. In 2016 I listened to so many of my apolitical friends ask, "How could anyone vote for Hillary Clinton, given how big of a liar she is?"

I couldn't believe how many people I heard say this. Hillary Clinton is a fine and hardworking person who has given much to this country and is not a

Figure 5-3. A great moment at the March for Science. I looked up and saw a group carrying these figureheads of science. Who should I see but Dr. Shirley Malcom, alongside the greats of science communication.

liar. She, like every other politician on the planet, has made some contradictory statements at a trivial level, but she has never been charged with any sort of perjury or other aspect of dishonesty.

And yet, as I've said, her team did a disastrous job of developing and communicating a clear story about her. I discussed this with Shirley in countless e-mails, and she kept saying that this is what happens when you don't manage to tell your own story well. Because of those e-mails, I decided that this principle needs to be called Shirley's Law.

Someone else—namely, Donald Trump—told Hillary's story for her and gave her the simple label of "Crooked Hillary." It was painful to watch.

I could give so many examples of Shirley's Law. It was at the core of much of what Michael Crichton addressed in his speech—scientists angry that they are misrepresented when in fact they fail to represent themselves. It's a cruel world out there; you have to figure out your narrative and then set about communicating it aggressively. This is what my Story Circles Narrative Training program is all about.

In the End, It's the "A" of the ABT That's the Secret to Listening

Like everything else, the challenge of listening tracks back to the ABT narrative template I am now rabidly promoting. The deeper I look into the ABT—more specifically, the three forces of agreement, contradiction, and consequence—the more I see the answers to all facets of communication. So let's now look at the "A" element in the search for how to improve as a listener. Let me recycle a little story I've been told before.

A woman told me that she and her husband had been invited to a dinner in Western Australia with a group of business leaders. She was seated next to the CEO of a coal-mining company. She began talking to him about climate change and the damage coal burning does to the atmosphere. Within a few minutes he was annoyed, was done talking to her, and directed himself toward the other end of the table for the rest of the evening. She asked how she might have handled it better.

The answer is right there in the ABT. The process of effective communication needs to begin with agreement. Before you jump into battle, first open up the lines of communication. The "A" of the ABT is for "And," which does exactly this—it is the most common word of agreement. For that to happen, you need to listen.

Imagine if she had started her conversation with him by chatting idly about sports in Australia. Imagine he had offered up that he's a mad keen golfer. Imagine she had listened closely, taken this in, and then disgorged every piece of knowledge she had of golf.

Through listening first and finding common ground of agreement, the narrative process can begin. Then, once a clear channel of communication is firmly established, it is finally time to introduce the source of contradiction—in this case, differences of opinion on coal mining.

Herein lies the very problem of literal-minded scientists, as I discussed in chapter 2. They see no reason to waste time going so far afield as to talk about golf when what's on their mind is climate change. They want to cut to the chase, jumping straight to the conflict. But once you've fired up the narrative part of someone's brain through contradiction, it's pretty hard to pull it back.

My advice for the woman in Australia was, basically, that you need to listen first to achieve effective communication. The question is whether this person who claims to be seeking advice really will listen once it's given. Sometimes scientists hear it, decide "Nah, that's not right," and go on about their way. But at other times, I swear they just can't help it. They're trying to listen, but their brains have been programmed, and here's what you get.

The Automatic Reset of the Programmed Brain

I really am not anti-scientist. It's easy to infer from all my criticisms that I just don't like scientists, but that's wrong. They are the most honest and best intentioned group of people I've ever met in all my travels. And some of them—it's almost heartbreaking when you realize it—are just stuck with the way their brain has been programmed. Like this person.

About the time of publication of the first edition of this book, I was near the end of the Shifting Baselines Ocean Media Project, which I talked about in chapter 1. A couple of groups in Los Angeles contacted me in desperation about the new Marine Protected Areas (MPAs) being created around 2008. MPAs are parts of the coast that are blocked off for "no fishing." It's always a heated issue because the fishermen don't like losing their favorite fishing grounds.

The conservation groups were desperate for some sort of media campaign. I set to work looking for an angle on how to address the issue, and I ended up talking with a friend who had been involved with the state-funded program to create the MPAs. We had a two-hour call, going through all aspects of the issue. As I searched for a theme, I began to hear her say one simple message repeatedly. She said, "What's a shame is that ten years ago we didn't have enough research on MPAs to say whether they were a good idea, but what people don't realize is that we can now show that MPAs work."

It was those last two words. I began saying them to myself over and over again. There was a ring to it—"MPAs Work." It was a potential slogan. It was great in that it was short, punchy, and very positive (about the solution, not the problem), and it didn't involve any blaming (not like "Fishermen wrecked our resources," which was the basic message of one group's failed campaign).

I put together a whole project and filmed a television ad that featured actor Pierce Brosnan (the media covered it as "James Bond Saves the Oceans!"), and the campaign proved successful in combating the forces of overfishing. As a result of the success and media coverage, I received a number of inquiries from other groups around the country wanting to run the same campaign.

So here's the sweet and even cute punch line. One scientist in Hawaii wrote a very nice e-mail saying he thought the campaign was excellent. He said it was exactly the message they needed over there for the same issue, BUT . . . (here we go with science-think) . . . he said they were thinking of modifying the slogan to say "MPAs *with Management* Work."

Ah, scientists. It's not that he didn't hear my short and punchy slogan. It's just that his brain was programmed to value information over concision. I'm sure he thought his twenty-two-character slogan was just as good as my eight-character slogan. I'm sure if I had shouted, "NO, THEY'RE NOT THE SAME!" at him, he would have been totally confused and said, "Geez, you don't have to get so snippy, Mr. Crankypants."

Some scientists just can't help it, but they really do mean well.

Listening Advice You Can Use: How to Present a Poster

Last bit here. In 2014, I brought a Groundlings improv actor named Samantha Jacks to the Ocean Sciences Meeting in Hawaii. She ran the improv part of my workshop. She had never been to a science meeting but is a bright, enthusiastic, and talented actor in her late twenties.

After our session finished, a student asked Samantha for advice on how to present a research poster at the meeting. Without hesitating—even though she hardly even knew what a research poster was—Samantha confidently explained exactly what to do.

She said, "Three things. First, you work out the ABT of your project. When someone walks up, you tell it to them. Then, second, you ask them what they work on, and, more important, you LISTEN as they answer. Then, third, you figure out what the bridge is that connects your work with theirs."

I was a scientist for about twenty years. I had never heard anyone explain how to present a poster at all, much less that elegantly. But she's absolutely right. Moreover, you can see the toughest part to making it work is the need to listen. But if you can pull it off, it's the secret to making your work of interest to everyone. You listen to them and they will listen to you.

CHAPTER 6

Be the Voice of Science!

In 2003 I filmed the Ocean Symphony *public service announcement, for which Jack Black agreed to be the conductor. I sent him an e-mail with pages of detailed notes on exactly what sorts of dance moves and wacky antics I needed from him and the specific topics in ocean conservation the PSA would address. At various places I asked him if he could tell a joke about this or that—you know, the standard one-liners about dinoflagellate blooms and anoxic events. In response to my several pages, he wrote back a single sentence: "I can't tell any jokes but I can conduct like a mofo."*

Once upon a time, communicating science to the general public was incredibly easy. In the 1850s Louis Agassiz, founder of the Museum of Comparative Zoology at Harvard University, gave public lectures on the Cambridge Common that were hugely popular. Hundreds of people would crowd around in the hot summer swelter to hear him speak for three to four hours—not telling religious stories (the heart), not performing stand-up comedy (the gut), and definitely not telling dirty stories (the lower organs). No, they would listen for hours as he talked about . . . fish taxonomy.

What was wrong with those people? I think they had empty heads. There was no television, no Internet, no iPhones, not even electricity. Their brains were in desperate need of stimulation. Think of it—just hearing words and information—it was like letting their brains sit in vibrating massage chairs for three hours. It must have felt sooo good.

Things have changed a bit since then.

What do you think would happen today if Karel Liem, a fish anatomist and brilliant lecturer at the same Museum of Comparative Zoology, started giving three-hour outdoor lectures on hot summer days about fish taxonomy?

The audience has changed. But it's not clear that the world of science understands this. Which is kind of surprising, since there is an entire field of science dedicated to the study of change called "evolution." But that knowledge seems to get focused more on the study of fossils than on the study of how the general public has evolved.

There exists today a new media environment. The large science organizations have been slow to adapt to it, but at the grassroots and individual levels things are different—change is indeed under way. You can see it in the proliferation of new modes of communication—from blogging to video-making to styles of graphic presentations. Individuals in the world of science are not waiting for the large organizations to show them the way; they are in the process of themselves becoming "the voice of science."

What exactly is the voice of science? At its very best, it was and still is Carl Sagan. In case you don't know, he was an astronomer, astrochemist, and unparalleled popularizer of science. He got his start in media fame with his best-selling book *The Dragons of Eden* in 1977 and was propelled to superstardom with the 1980 television series *Cosmos: A Personal Voyage*. Given the scale of achievement of his popular books and television series, he is the most successful scientist in recent decades in communicating pure science to the general public. So let's take a look at him now in relation to the previous four chapter titles.

First off, he was certainly cerebral (chapter 1)—a truly great thinker. *However*, he was not so caught up in his thought processes that he was unable to act. In fact, as William Poundstone mentions in his biography of Sagan, he acted so

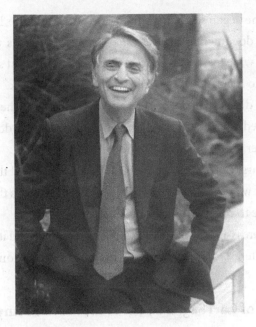

Figure 6-1. Carl Sagan, proof of how not being "such a scientist" can help you connect with the general public. Photo by Michael J. Okoniewski.

vigorously in his efforts to communicate and popularize science that his friends often wondered when he slept.

Carl Sagan also was not constrained by being overly literal-minded (chapter 2). How else would he have ended up on Johnny Carson's *Tonight Show* couch so many nights? Many scientists would have considered the show a bunch of silly banter, but Sagan fully understood the enormous power of television to influence American society.

How about storytelling (chapter 3)? Sagan's 1985 novel, *Contact*, was a best seller and was made into a major motion picture. The man appreciated the power of telling a good story, and, while not religious himself, he had a tremendous grasp of the role of religion and mythology in the human psyche. It was a theme he explored in many of his books, beginning with his first best seller, *The Dragons of Eden*.

Likeability (chapter 4)? Sagan was dorky, nerdy, and even goofy at times, but he had the "it" factor that made him very well liked. In the late 1990s a friend of mine, filmmaker Mark Shelley of Sea Studios, got to see Carl Sagan's lasting influence when he began searching for an on-camera host for his National Geographic Society documentary series *The Shape of Life*. His production team searched far and wide for a host: they auditioned a number of scientists, showed the test footage to focus groups, and listened to the feedback.

The viewers didn't like any of the candidates. It drove the producers crazy. They finally asked the focus group members, "Well, who do you want for a host?" The answer was very simple: "Another Carl Sagan."

Over the course of several decades Carl Sagan embodied the very best traits of a scientist and was widely loved as a result. Overall, he was the living proof of everything I have to offer with this book.

However, there is a sad footnote to Carl Sagan's career, and it's something that every scientist interested in engaging in broad communication needs to know.

The Rejection of Carl Sagan by the National Academy of Sciences

A number of fine biographies of Carl Sagan recount the unfortunate details of his treatment by the National Academy of Sciences.

In a nutshell, Stanley Miller (of Miller and Urey fame, the team who were among the first to describe mechanisms for the possible abiotic origin of life on Earth) headed up a group that nominated Sagan for admission into the National Academy of Sciences. The Academy is sort of the equivalent of the Hall of Fame for football and baseball players at the end of their careers.

Sagan made the cut in the initial voting, ending up in the top 60 of the 120 nominees that year. This was enough to secure his entry into the Academy, provided no one objected to his induction. Of the previous 1,000 nominees who had made the cut, only one had been objected to by any member. Sagan ended up being the second. Which meant there had to be a special vote for him.

Before the vote, there was an open debate in which many members lashed out, denigrating him for supposedly being a lightweight scientist despite having published more than 100 peer-reviewed papers and numerous books and having made major accomplishments in astronomy. Texas A&M University chemist F. Albert Cotton referred to Sagan's involvement in the popularization of science as "symptomatic of an inadequacy in *doing* science."

In the end, Sagan needed a two-thirds majority for admission but failed to get it. How in the world did this happen?

Sagan's first wife, Lynn Margulis, who had an unhappy divorce from him (I once heard her lovingly refer to him as an "ass" in the question-and-answer session of a talk she gave at Harvard), was part of the team fighting for his acceptance. She set aside all personal animosity and vigorously defended his distinguished career.

According to biographer Poundstone, Sagan never spoke with any bitterness publicly about the defeat, but in a letter, Margulis told Carl that Cotton's speech

> resonated with every small mind, ugly body, and verbal maladapt present, and that means half of the membership. They are jealous of your communication skills, charm, good looks, outspoken attitude, especially on nuclear winter. . . . In summary, you deserved election to the National Academy years ago and still do; it is the worst of human frailties that keeps you out: jealousy.

In short, the Academy never forgave Sagan for being so popular.

That one event stands as a monument to the risks of broad communication. And while today more scientists than ever are involved in communicating science, and the National Science Foundation even requires recipients of its grants to set aside a substantial part of the funds for "outreach," the dilemma still exists. In fact, I'm even willing to put a number on it.

The One-Third Rule of Science Cannibalism

I'm gonna go ahead here and propose the hypothesis that, all else equal, in any random group of scientists, about one-third of them will simply dislike anyone who stands above the pack and tries to communicate directly to the general public. If you are a scientist, I invite you to put this hypothesis to the test.

I base this number first on the fact that a third of the Academy voted against Sagan. But I further support it with a few personal experiences.

At Scripps Institution of Oceanography, for several years I taught a week of communications during their twelve-week intensive orientation course for new ocean science graduate students. Each year at the end of the course, the students would fill out evaluations. And each year, the same simple pattern emerged: about a third of the students talked about my communications section as a "life-altering" experience. They so thoroughly enjoyed themselves and found it so eye-opening that they felt certain a major part of their future scientific work would involve communications projects.

A second third called the week very worthwhile. But the final third, oh, yeah. You guessed it. They lashed out, calling it a total waste of their time, insulting me as some sort of "poser," seeing no relevance of the material to their career in science, and basically hinting at the possibility of a refund of their tuition.

The sentiments of the students, each year, matched the reviews for my movie *Sizzle: A Global Warming Comedy*. About a third of the scientist blogger reviews raved about the movie, a third called it adequate, and a very vocal final third seemed to be channeling the voices of the National Academy that voted down Sagan.

Uncritical Science: Like an Ocean with No Sharks

You could also see the one-third rule when, in late 2007, a novel idea emerged to organize a debate among the U.S. presidential candidates focusing on scientific issues. Termed "Science Debate 2008," the idea began with Matthew Chapman, the great-great-grandson of Charles Darwin, who decided he was fed up with the amount of attention paid to religion in the presidential debates in contrast with the virtual absence of science discussion. He managed to create a groundswell of support for the idea, eventually recruiting all the major science organizations and a gaggle of Nobel laureates to back it. But along the way, had you read the science blogs, you would have found rather consistently about a third of the voices negating the entire idea, saying it wouldn't work for a variety of reasons, despite the endorsement by most of the top leaders of the science world. Most of the skeptics seemed to have some basic aversion to getting involved in politics and engaging with the general public. Or they were just contrarians by nature.

So the negative, negating, cynical voice of the science world is still there, and it will probably be there for the rest of time. Which is fine. You wouldn't want it to go away. If it ever did, there would be reason to worry. It would be like an ocean with no sharks. As much as you don't want to get attacked by a shark, you also don't want them to go extinct. With some things in life you just need to find a way to coexist.

Personally, I sensed this problem long ago. I knew from the start of my media involvement that I didn't want to end up as an academic attempting to do both serious research and silly filmmaking. It's just too big of a divide. I might have been able to make it work, but it would have been an unfair imposition to make on my colleagues and students—to ask them to accept me as both authority figure and clown.

In fact, while I was still a professor at the University of New Hampshire, I was so keenly aware of it that for my first short film, *Lobstahs*, I used the fake name Charlie Agassiz for the credits. There was no trace of my real name. Of course, I eventually dealt with the dilemma by resigning from my professorship. But that was because I sensed I would be going much further into the media realm than would a normal academic scientist.

As for the central divide between the purists and those who would deign to reach a hand out to broader audiences, you can see pretty much the same dynamic in other professions. Take, for example, the life of John Steinbeck. His literature was enormously popular with the average reader, but he was forever dogged by the critics. When he was finally awarded the Nobel Prize in Literature, there was such an outcry from the literary world that at his press conference a reporter actually asked him whether he thought he deserved it (and, sadly, he humbly replied that he wasn't actually certain he did). Had there been a National Academy of Sciences for him to be elected to, he probably would have suffered the same fate as Carl Sagan.

Yet when all is said and done, in the world of science, criticism is an essential part of the process. There's no denying it. And it leads us back, one last time, to my beloved acting teacher.

The Delicate Art of Negation

The woman was mean. Make no mistake. I could tell you some truly horrible stories about her—like the night she got into a fight with a student who had criticized her teaching style. She accused the student of acting out with her the same mother-daughter problems the student must have experienced growing up.

The spat quickly escalated until the student burst into tears and ran out of the classroom, followed by the instructor. We all sat listening to the screaming match that ensued in the hallway, which culminated with the student shrieking, "F*** you, don't you talk about my mother, she died five years ago!" and the teacher screaming back, "Good, well, why don't you go visit her grave and take a sh*t on it!"

Honest to goodness. Doesn't that make you want to go take an acting class? Hollywood can be such a sweet and nurturing place.

But every third night we had a different instructor at the school. He was a very nice man—handsome, friendly, never confrontational, always supportive, allowed everyone to talk, always listened. And for the first month everyone craved the night with him, desperate to get away from the old bag.

But guess what eventually happened? Everyone got tired of the nice guy. After a while, all his sweetness and positivity was okay; it just wasn't very interesting, exciting, or challenging. There was an electricity to going into class with the mean woman. Everyone sat on the edge of their seats, wanting to do their best work. She was tough, she was critical, she was brutal, but when she finally said something was good, it really meant something.

She was terrible to me, and she doesn't like me to this day, but I'm forced to concede she was an excellent instructor. The bottom line is, let's face it, really good teaching sooner or later involves a certain amount of pain. Whether it's the pain of tedious work or the pain of stinging criticism, seeking a totally pain-free education is kind of wishful thinking.

It's important for young scientists in particular to take note of these dynamics. It is essential for them to become better communicators by learning from, rather than being crushed by, criticism. And also, unfortunately, to expect some lack of interest in communication from the science world. Especially since there is one very important large-scale pattern that I have observed over the course of my twenty years of exploring science communication.

Find the Good Communicators!

As the years floated by, I hit countless brick walls in seeking support for innovative ways to communicate science. (A program officer at the National Science Foundation said, "Scientists are pretty much just going to communicate the way they always have—the good ones will always be good; the bad ones will always be bad.") But I also found a number of very firm and committed backers. Eventually I hit a point where an unmistakable pattern emerged.

Here's the pattern: Good communicators believe in the power of communication. Poor communicators don't.

More often than not, when you encounter scientists, administrators, foundation officers, or even politicians who feel that spending good money on such stupid things as television commercials or movies is a waste of limited resources, you will find that they themselves have relatively poor communication skills.

I think there is a positive feedback loop—or, in less sciencey terms, a snowball effect—that develops throughout people's lives. Those with poor "people skills" have had a lifetime of disappointment with using communication (i.e., persuasion) to get their way. They try to speak to their next-door neighbors about keeping their incessantly barking dogs inside at night but find they can't persuade them through discussion. So eventually they give up on that approach and instead go straight to the police and get action through a more objective means—force of law.

The more this happens, the more they conclude that talk is a waste of time— "Let's 'do' something" becomes their attitude. If they end up being in charge of a conservation group, they'd rather bring in the lawyers and propose legislation to stop the new housing development next to a park than launch a mass media campaign to convince the local residents to defend their natural resources. When somebody stands up at a meeting and suggests spending money on better communication, they respond by claiming they already have and it didn't work.

Conversely, there are others who naturally have strong "people skills." They find it very easy to speak to their next-door neighbors, have them over for a beer, laugh about the barking dogs, and then make a friendly deal to give them their extra lawn chairs in return for keeping the pets in at night. The law is never needed. Everything is settled through the more subjective means of communication.

Years later, they have a natural appreciation for the importance of communication. They've used it effectively throughout their lives. When you say, "We need to put more effort and resources into communication," they respond positively because it has worked for them so well in the past.

This has certainly been my experience. *All* of the scientists who have supported my efforts have, not by coincidence, themselves been excellent communicators. So my simple advice is to seek support from these people—those who know how to talk and listen. They already understand the power of good communication. And they can help you with the most important aspect of your communication effort—to discover and nurture your "voice."

Which brings us to a very important term in communication—"voice." It means much more than the sounds emerging from the vocal cords. Whether it's the "voice of authority" in warning signs at the airport or the voice of compassion behind humanitarian efforts, this subjective element is central to effective communication.

The Scientist's Voice

Yes, that was me back at the beginning of chapter 1 with the letter to the editor of *Premiere* magazine scolding Marky Mark Wahlberg for misidentifying whales. By now, I know my voice all too well, and I know that despite nearly two decades of running around Hollywood, hanging out with actors, making movies, and trying to pretend I'm a Hollywood player . . . I still have the voice of a scientist. And that's fine.

Whether it's genetic or developmental, who knows, but surely it's there for life by the time you've completed a doctorate in science. I'm stuck with it.

Here's a little example of my public voice. At our orientation day lecture at the USC School of Cinematic Arts, all fifty members of our class were seated in a theater and had to introduce ourselves with a few sentences. When it came to me, I said, "I was a marine biologist. I earned my Ph.D., spent a number of years in Australia working on the Great Barrier Reef, and—" To which a professor interjected, "Did you fix it?"

I didn't even understand his little quip until later. In the world of science, one of the first things you ask a colleague is "What are you working on?" or "What are you looking at?"—as in "I'm looking at speciation in sympatric populations of hoppy toads." If you're a scientist, this is just commonplace dialogue. But to that professor of filmmaking it sounded as if I were some sort of repairman out in nature working on the reef.

The bottom line: There is a science dialect that you pick up without even realizing it. You think you're talking "normal," but the civilians hear otherwise. This is one of the hardest things to get a lot of scientists to realize—they feel like, "What do you mean? We talk just the same as everyone else." And they're certain of this (with a probability of error of less than zero point zero five).

It certainly has taken me a lot of years to fully realize the differences. There are benefits to having the voice of a scientist (some people are actually impressed!), and there are costs. And in my case the cost is even financial.

The Life That Didn't Happen

In 1990—that's *nineteen ninety*, mind you—getting close to three decades ago—I wrote a novel titled *Ice Blue*, which was a tale of suspense in Antarctica drawn from my experiences there in the mid-1980s. It was a wild and woolly story involving shipwrecked oil tankers, diving beneath the Antarctic ice, and frantic helicopter chases over the frozen wilderness.

Though I was still a professor at UNH, I had a celebrity friend who had a literary agent at William Morris, one of the big three agencies in Hollywood, to whom she gave the book. He was an avid scuba diver, he read it, he loved it, and he called me up asking to represent it. A month later, I was meeting with him in Hollywood and he was asking me whether we should go for a $200,000 or a $300,000 advance and which A-list actors should play the lead roles in the movie version. I did my best not to be snowed by the Hollywood hype.

In an alternate universe, that book got bought by a leading publishing house, spent a year on the *New York Times* best-seller list, and became a huge blockbuster movie that made me a gazillionaire and put me in the same club with Michael Crichton.

Unfortunately, in the universe in which we live, the book was turned down by a dozen publishers. My agent got rejection letters from the same editors whose names I saw in best-selling books by Tom Clancy, Scott Turow, and John Grisham. The book was being read by the best sources, but there was something wrong with it.

Today I can look back and tell you fairly precisely what the problem was. The narrative voice was the voice of a scientist. It was very honest, humble, accurate, and precise. All of which meant it wasn't the best fiction storytelling. The entire novel was woven out of real tales that I'd heard when I was in Antarctica doing research. In the climax scene, the heroine swims beneath the ice in freezing water that is nevertheless about fifty degrees warmer than the air temperature. Her soaking wet attacker tries to chase her but finally freezes to death when he climbs out of the water and into the frigid air. My agent loved that scene and never got over it—he still talked about it years later, asking me repeatedly if it was true that the air would be so much colder than the water, which of course it is.

But for the editors the story just wasn't big enough, grand enough, and sensational enough. I'm guessing they felt it needed nuclear weapons, abominable snowmen, or space aliens (as my writer friends suggested years later).

The book never got bought, but the scientist voice stayed with me. To use a line that occurs in both *Dodos* and *Sizzle*, I was, in the end, "handicapped by a blind obsession with the truth."

This is the burden that scientists and science communicators face. It is the eternal struggle between storytelling and reality, which I hope I've made clear in this book. It is not an insurmountable challenge, and when a good story can be told using completely accurate facts from the real world, that is often the most powerful story of all.

Many of the greatest movies ever, from *Lawrence of Arabia* to *Titanic*, are woven primarily out of the truth. Liberties are taken to make the stories work to their maximum strength, but in general both films reflect an effort to stick to the truth.

So I am in no way encouraging anyone to distort science, only encouraging scientists to help the rest of the world understand the crucial insights of their work by going the extra mile. Scientists actually know what I'm talking about here. They write research papers. They know what it's like to take a rough first draft and turn it into a well-polished finished product. The only question is how far you are willing to push. And this comes down to the importance you place on communication.

Let the Revolution Begin: Dada and Attention

Now it's time to look to the world of art and propose a parallel pattern for the world of science, separated by roughly a century. In many fields of science, there is a progression from an old-fashioned *descriptive* phase to an exciting new *experimental* phase.

In my field of marine invertebrate embryology, for at least a century scientists sat at their microscopes and humbly described everything they could see when they looked at animal embryos in various stages of development. But then a lot of scientists seemed to tire of simply describing what they saw and began wanting to know how things work. This led them to begin experimenting. They began turning their efforts toward subjecting embryos to different physical conditions or removing individual cells and watching what happened. Instead of just describing patterns, they began to delve deeper into understanding the processes that create the patterns.

The "descriptivists" (if we can coin such a term) are generally seen as traditional, conservative, unimaginative, content with the status quo, and even repressive toward those who challenge their established ways. The experimentalists are regarded as more bold, brave, confrontational, innovative, and determined to break free of the bonds of tradition. The descriptivists are also seen as more disciplined, while the experimentalists seem more reckless.

Figure 6-2. *Fountain,* the piece of art with which Marcel Duchamp shocked and offended the conservative art world in 1917. Could the science communication world use a similar shake-up?

In the world of art, a major transition of this sort began with a single, explosive event in 1917 staged by the playful French artist Marcel Duchamp. In an era when art was considered to be grandiose paintings by the great masters, Duchamp and two friends purchased an iron urinal, turned it on its side, called it *Fountain,* and submitted it to an art show in Zurich, Switzerland. The judges, being traditionalists, deemed it "not art" and rejected it. An uproar ensued among the newly formed Dada cultural movement (which produced what is sometimes called "anti-art" in a fashion similar to antiplot, discussed in chapter 3). Duchamp protested by resigning from the board of the Society of Independent Artists, and the art world was never the same.

A full appreciation of this event requires an understanding of the social tensions in Europe on the eve of World War I and the widespread frustration with the staid and serious tone of the art world in general. The urinal was a statement of rebellion, which was met with hostility by many established artists yet laid the groundwork for much of today's modern art movement. To this day *Fountain* is pointed to as a landmark creation—so much so that it was the centerpiece of an entire museum show in 1989.

The discussion of "what is art" continues, and the bewilderment of so many when looking at what Duchamp referred to as "readymades" (objects that become art simply by being labeled as such) has not abated. In 2007 a wonderful documentary, *My Kid Could Paint That*, took yet another look at this issue by asking whether the abstract paintings of a child could really be considered art. But there is a different way to view the work of such artist provocateurs.

Richard Lanham, in his discussions of what he calls our "attention economy," offers up a new and interesting perspective on Duchamp's defiant gesture. If you view Dadaists and their descendants, such as Andy Warhol and Christo, not so much as artists but rather as "attention economists," you'll see they were brilliant. They knew how to work within an economy in which attention is increasingly the most important currency, and they knew how to score big.

All of which lays the groundwork for what we now see in the communication of science. Today there exists a very traditional style of communicating science. It is quiet, reverential, rich in information, supremely accurate, short on humor, devoid of emotion, and increasingly ineffective, as I noted in the discussion of the first of Laurie David's two global warming movies.

In the world of documentary filmmaking in general, the equivalent of *Fountain* could well have been Michael Moore's 1989 groundbreaking film *Roger and Me*. In the film, Moore dispensed with the age-old ethic of filmmaker as outside observer or "documentarian" and instead jumped in front of the camera as an active participant. Documentary purists were as angered by his work as the art purists had been by *Fountain*. But it also became the most commercially successful documentary in history at the time.

In the same way that Duchamp caused the art world to ponder "What exactly is art?" Moore prompted the documentary filmmaking world to consider "What exactly is a documentary?"

And now, in the postmodern era, the presentation of science through the traditionally rigid channels of science documentaries seems to invite experimentation. And that, in essence, is what Laurie David and Al Gore did—*An*

Inconvenient Truth was experimental and also drew on Gore's celebrity to grab attention.

The bottom line: experimentation is essential to advancement, and to experiment successfully you must be able to draw on your voice.

Your Voice

Everyone has an expressive, creative, personal voice to some degree, and it comes with a stamp that is as indelible as fingerprints. We saw this in film school. In our first class, we had to make five short Super 8 films. Each week, five of the fifteen students would show their latest "masterpiece." In the fourth week the instructor took one week's films, chopped off their opening credits so no one would know who the filmmaker was, and then after each film asked everyone to guess who made it.

It was so easy. Everyone was able to guess: "Oh, that was a Javier film," "Oh, that was a Zellie film," "Oh, that was an Ann film." Everyone's style had become so obvious that even if the guy who had made three Tarantino-like guns-and-violence films were to switch to a romantic comedy, you'd still sense his style. Something would just come through. That something is called "voice."

If you are a scientist (or pretty much anyone) and you want to get involved in mass communication, this is your official starting point. You need to listen to your "voice" and figure out what it is.

I know what mine is. You can see it in my twenty years of filmmaking and even before that. It's characterized by bright colors, upbeat music, silly and even campy humor, "high key" lighting (meaning brightly lit scenes, as opposed to dark and moody), simple but well-structured stories, sincerity, irreverence bordering on offensiveness, and a tendency toward provocation.

Spending three intense years in film school had no effect whatsoever on my voice. I made my barnacle music video in 1991 and *Sizzle* in 2007. They were sixteen years apart, with film school in the middle, but for better or worse they have the exact same voice—same combination of silly and serious, science and nonsense, bright color, lively music, and plenty of humor.

And I can name for you some elements my voice doesn't have—non-melodic music, complex and inverted story lines, amazing visual gymnastics, large doses of mystique, eyeball-wrenching rapid-cut sequences, tons of beautiful people, on and on. That's just not part of my voice.

So when it comes to communicating with the general public, what is the voice of science today?

First off, as with any discipline, the strongest voice is that of a single individual. When the United States goes to war, it starts not with a press release from the Pentagon but with a single individual, the president, standing before Congress and lending his single human voice to the mission.

There is nothing more powerful than the first-person narrative—the voice that can speak to a crowd and say, "This is what I know; this is what I have experienced; this is how I feel."

In 1999, four authors published *The Cluetrain Manifesto: The End of Business as Usual*, which was an essay directed at the business community that underscored the importance of the human voice by demanding respect for the channels of communication opened up by the newly created Internet. They offered up the following warning regarding the markets available via the Internet:

> These markets are conversations. Their members communicate in language that is natural, open, honest, direct, funny, and often shocking. Whether explaining or complaining, joking or serious, the human voice is unmistakably genuine. It can't be faked.
>
> Most corporations, on the other hand, only know how to talk in the soothing, humorless monotone of the mission statement, marketing brochure, and your-call-is-important-to-us busy signal. Same old tone, same old lies. No wonder networked markets have no respect for companies unable or unwilling to speak as they do.
>
> But learning to speak in a human voice is not some trick, nor will corporations convince us they are human with lip service about "listening to customers." They will only sound human when they empower real human beings to speak on their behalf.

Whether the *Cluetrain Manifesto* has been of any importance to the Internet is debatable, but regardless, the basic message is completely relevant to the communication of science. It is about fostering the individual human voice, and this is something that the science world has been slowly coming to realize.

In my thirty-some years of following the science community, I have seen individuality and the progressive "humanization" of science slowly emerge. When I was an undergraduate in the 1970s, there were still remnants of the robotic, disconnected, third-person style of presenting science in both writing and speaking that had been perfected in the post–World War II era of research. Scientific research papers would use this weird, otherworldly voice, saying, "The investigator collected samples," even though the author *was* the investigator. That's like Mr. T saying, "When Mr. T wants to whup a**, Mr. T whups a**."

That doesn't go on much anymore. Most journals today allow you to speak more directly and just say, "I collected samples." And public speaking is far less stilted and formal than it was decades ago.

In terms of compassion (another human element), when I was a graduate student, young scientists finished their doctorates and were coldly dumped into the marketplace with an air of Darwinian models of natural selection. The most miserable and embittering five years of my life were from 1983 to 1988 as I desperately sent off job applications in search of a tenure-track professorship. During those years, senior professors told me "Good people get jobs," meaning "There's nothing you can do; the job market is going to tell you if you're a good person."

It's still a tough process today for young scientists finding employment, but at least several decades' worth of editorials in *Science* and *Nature* have awakened some level of compassion. Today there is a Naturejobs website to assist struggling young scientists, and most science organizations invest a lot of effort in helping fledgling scientists find their way, with, it is hoped, a little less heartache than my generation experienced.

The world of science is slowly, gradually becoming more humanized. And that allows room for individual scientists to increasingly speak out in their own distinct voices.

Which brings us back to the individuals from the science world who feel drawn to engage with the general public. Yes, I'm speaking to you—the scientist, the science communicator, and even just the science aficionado. I want to finish by offering up a few words of encouragement, as well as reality, when it comes to communicating science to the general public. Let me start by ground-truthing this book.

Is This Book "the Be-All and End-All" for Communicating Science?

So what were you expecting—that the book would be the definitive manual for communicating science to the public? Here's one more Hollywood anecdote.

In the first summer of film school, I got a job working as an assistant on the Hollywood movie *Three Wishes*, starring Patrick Swayze and directed by Martha Coolidge. It was great (if you rent the DVD you can see my name in the end credits, buried among the production assistants and listed as Dr. Randy Olson). I was thirty-eight years old, a tenured professor (technically I still hadn't resigned), and there I was getting coffee and lunch for all the producers, many of whom were younger than me. It was wonderfully humbling!

For a couple of weeks I managed to get assigned to work with the casting director, videotaping auditions. As a number of major actors came in to "read" for parts, I sometimes got to chat with them in the waiting room. I had just begun my Meisner acting course, which, despite my beatings, I was very excited about, so I would ask them if they had ever taken such a course and, more important, what they thought was the best and most essential acting class to take.

What I heard from every single actor was the same thing. There is no one class or method that is the definitive education for acting. Instead, you need to take a variety of courses—scene study, cold reading, improvisation, Meisner, and so on—and take what is of relevance and value to you from each one, in hopes that together they will eventually make you into a well-rounded actor.

I give the same answer for science communication in relation to this book. As I mentioned in chapter 1, a number of excellent workshops and guidebooks are now being offered that address how to mold scientific information into understandable messages. That knowledge is essential to communicating science effectively, and I haven't even begun to delve into it here.

This book alone is not designed to train you as a mass communicator. It's more of a lesson in how to rethink your style of communication so that you can reach a larger audience.

So Then, What Does the Title Mean?

Now that we've almost completed our journey, let me step back and address the title of this book, which does not say "Don't *be* a scientist." It merely says don't be *such* a scientist.

I had an incredible amount of fun in all facets of my career as a scientist (except for writing grant proposals—ugh). I loved doing research, loved going to scientific meetings and giving talks, loved reading (good) scientific papers, loved writing research papers that got accepted (not quite as fond of the ones that got rejected), and more than anything else loved the application of reason and logic to the natural world through the scientific method. I spent a year on Lizard Island, Australia, living and breathing science every single day. It was the very best year of my life.

Some day I will make it back to Lizard Island to resume my research on the strange little white and brown blobs that I studied for my doctoral dissertation. My heart will forever reside in the Lizard Island lagoon. I loved my career in science and departed only because I equally enjoy the telling of stories through film. There is nothing negative about a career in science intended by the title or anything in this book.

No, the fact is, the title of the book is *Don't Be* Such *a Scientist*, and I will now, reluctantly, reveal where the title comes from. I spent eleven years, the better part and certainly the best years of my science career, with a woman whom I married and eventually divorced. She was not a scientist by any stretch. In high school she was an accomplished singer, dancer, and actor who could easily have pursued a career in those professions but instead chose to study environmental policy, eventually earning her master's degree.

Throughout the years we were together she was my biggest fan and supporter, going to scientific meetings with me and spending months in the field with me at marine biological laboratories and countless nights listening to scientists do what they like to do most—talk about their research. She was one of the all-time greatest fans of science and an incredibly good sport. And yet . . .

There were times when being around the science life was just too much for her. The spouse of any scientist knows what I'm talking about, in the same way that the spouse of any lawyer, accountant, politician, engineer, real estate agent, or most any profession knows (just take, for example, my wonderful mother— my father could easily have written a book about her titled *Don't Be Such a Real Estate Broker*). All of these professions can require a great deal of focus, intensity, and concentration, which leads to occasional phases of myopia. And because science is so information intensive, I would suggest that it might even lead to a little bit more of it than other professions.

And so, despite how much fun and fascinating the science life was, there were times when I would show some of the worst traits I've talked about in this book—the tendency to be so cerebral (preferring to read a book rather than go dancing), so literal-minded (unwilling to suspend disbelief and roll with a silly story that pushed the limits of credibility), and such a poor storyteller (going on and on about a scientific study because I believed the data to be so fascinating), and, in the end . . . at times being unlikeable (including moments of extreme cynicism). All of which did, on more than one occasion, lead her to cry out, sometimes humorously, but also sometimes with tears in her eyes, "Please . . . don't be *such* a scientist!"

This is what the "Don't be *such* a scientist" admonishment is about. Not to be any less of a scientist than your mind tells you to be, but simply to develop an awareness of where the excess focus will take you. You want a healthy, productive life as a scientist? You've got to find ways to develop an awareness of the myopic drive and the need to split your attention. In essence, you need to . . .

Be Bilingual

This is my specific recommendation. Know that there are two audiences for you as a scientist. I have talked about them both. Let me summarize here some of the basic dynamics in a form that scientists can relate to—a table.

Table 6-1. How the Broad versus Academic Audiences Respond to Various Aspects of Communication

	Broad	Academic
Main information channel	Visual	Audio and visual
Structure	Need a story	Information is fine
Mode of response	Visceral	Cerebral
Need humor?	Pretty much	Not necessarily
Like sincerity?	Always	Suspicious of it
Sex appeal?	The ultimate	Potential disaster
Prearoused?	No	Yes
Effective elements	Humor, sincerity, sex	Information
Effective organs	Heart, gut, gonads	Head
Preferred voice	Human	Robotic

As a scientist or science communicator, you need to become "bilingual"—to be conversant in your area of specialty in both languages.

There are, of course, plenty of exceptions. Many researchers are shielded from the general public and don't have to bother with the broad language, and many science communicators never have to come into contact with scientists, so they need to worry only about broad communication. Still, working toward bilingualism is a potential bonus for anyone associated with the world of science.

So how does this work? It means speaking the right language to the right audience. One of my best scientist friends complained to me recently about her graduate students giving talks at scientific meetings in which they show funny cartoons all through their presentations, to the point where the serious scientists find it annoying and even discrediting. It is. I've seen it with other scientists. As bad as it is for a scientist to speak with molasses-thick jargon to the general public, it's equally bad to speak with a broad, elementary voice to fellow scientists.

Truly Be the Voice of Science

If you want to make a major contribution to science communication, you need to know from the outset that it will be a long and personal journey. It won't be easy. It won't be safe. And it's doubtful you'll be able to control the timeline.

No one told Carl Sagan to write science fiction novels, get involved with Hollywood filmmaking, or go on Johnny Carson's *Tonight Show*. He simply had an inner voice driving him to reach out and share his passion for science. He *was* the voice of science, by his own doing.

In working on this book, I managed to contact his last wife, Ann Druyan. She said that while the National Academy incident was definitely a setback for Carl, in his final years he was more satisfied than ever before and thoroughly relished the joys of sharing science with enormous audiences around the world. He died a very happy and content man.

On a much smaller scale, I can offer up the same overall report. No one in the science world ever recommended my initial involvement in filmmaking: it all came from inside. I enjoyed connecting with broad audiences through film, and I began to experiment.

I didn't have any clear timeline. A month after I got to Hollywood, in 1994, I was at an entertainment industry cocktail party, standing in the buffet line and talking to a haggard old man who was a veteran agent. I was telling him about all my grand ambitions for filmmaking. While plucking hors d'oeuvres onto his plate, and without even looking up, he said, "So how long ya gonna give it?"

I didn't even know what to say. I'd bought a one-way ticket to Los Angeles. It never even crossed my mind to do anything other than this for the rest of time.

Fifteen years later, I am still making science-related films and can say that, on the whole, it is thoroughly rewarding. One of the most heartwarming experiences has been the response I received to *Flock of Dodos*.

Although none of the major science organizations showed any willingness to support what I had done, at the grassroots level an incredible number of old friends and new friends suddenly emerged. They contacted me about screening the movie at their universities, and what ensued over the next two years was a string of more than fifty major events with excellent panel discussions. None of it involved large science organizations.

All of which showed me that there is a new interest in the broad communication of science and that the greatest support for this movement resides in individuals.

Naomi Oreskes, star of my movie *Sizzle*, talks about how a hundred years ago scientists were by tradition very good at speaking to the lay public, as well as personally and passionately committed to do so. But that changed in the United States after World War II. The government began establishing enormous science agencies and programs and creating a new breed of research scientist who no longer needed to appeal to the public for support. A new standard emerged in which these scientists felt entitled to the right to conduct research without having to explain it to average folk. The heads of science organizations acceded to these desires of scientists, and the idea of communicating science to the public was shifted from second nature to a secondary priority.

Today, however, a change is in the air. Just as the Internet has revolutionized the individualistic drive of the population in general, it is also fostering a grassroots strength in the world of science communication. And with today's new individual science communicators comes a bit of rebelliousness. They are exploring new ground, pushing back the boundaries, and overturning traditions. They are tossing out the old and bringing in the new. But as they go their way, building these bridges to the broader lay audience, it's my hope that at least a few of them will follow a very simple rule of thumb. Perhaps occasionally, when they're not quite managing to connect with the public, one science communicator will whisper to another, "Maybe try to not be *such* a scientist."

Therefore: Update for the Second Edition

In the nine years since I wrote this chapter, a lot of people have gotten interested in the communication of science. At least three equally good books came out around 2009 along with mine—*Escape from the Ivory Tower* by Nancy Baron, *Am I Making Myself Clear?* by Cornelia Dean, and *Unscientific America* by Chris Mooney and Sheril Kershenbaum. Together, we provided a boost not only to the growing awareness that communication is an important part of science but also to the eventual realization that it's at least half of the whole enchilada. If a tree falls in the forest but the news of its doing so isn't communicated accurately, people can end up thinking it actually blossomed in the forest. What good is that?

But I want to rain on one little popular myth that will probably further reduce my popularity with academics, which is the idea that "younger people are better at communication." No, they're not. Not in my opinion. Yes, they are increasingly obsessed with devices to communicate, but if people are communicating poorly, are they really communicating?

I think we've seen this problem with my Story Circles Narrative Training program. The training works great for professional scientists but not nearly so well for graduate students. Why is this?

We've boiled it down to the fact that the participants have to have either a *need* or a *want* for the training. What we get with professionals is they have an immediate need for the training. They have witnessed countless projects that have failed as a result of poor communication, and their jobs can depend on their ability to communicate. The training is fairly challenging and needs a full head of steam to work properly. They have the motivation because they have a clear need.

But graduate students just don't know yet that they will need it. They tend to be overbooked, they haven't had that much experience with failed communication, and, most concerning, they do tend to think they are great at communicating because they are on Facebook and Twitter all day. But if everything you're posting and tweeting is non-narrative "And, And, And" material, you really aren't communicating all that much.

So it's a challenge. But that's okay. I like the fight. I've always liked the fight. In fact, let me close by sharing a few final personal details on my love of the fight, now that I'm sixty-two and increasingly opting for surfing over punching brick walls.

Fight the Power

If I'm lucky, in the next few years I will bring my family's World War II story to the big screen (or, more likely, to a television miniseries). My father was the last surviving officer of the infamous Bataan Death March and wrote three books about it; my grandfather was chief of staff for Douglas MacArthur at the end of World War II. If I do, you'll learn about the uncle I never got to meet. I was named after him (my first name, Richard, which I used until college, was in tribute to him).

Figure 6-3. My uncle the rebel, Richard J. Marshall Jr., 1942.

He was a troublemaker, forever fighting authority, and had a clear science bent. As a child he collected insects in the jungles of the Philippines. My mother—his kid sister—was his six-year-old field assistant. When I was in sixth grade she gave me his dissection kit, but I was too clueless to appreciate it.

As World War II began, my uncle enrolled at VMI, Virginia Military Institute, following in his father's footsteps. But he had immediate problems with authority, eventually went AWOL, was caught and thrown out, enlisted in the army, was sent to Europe to fight, and then was killed when his ship, the USS *Dorchester*, was torpedoed by a German U-boat off Greenland.

One time when I was in high school, my mother showed me a letter I had written to her that ended with "Love, Richard" and a postscript that said I would be home for the Fourth of July. The letter was accompanied with little artwork doodles of cannons and fireworks. It was in my handwriting, and those were my typical sort of doodles. She asked if I remembered writing it. I stared at it but couldn't remember.

Then she unfolded the top of the letter—it was dated June 1942. It was from my uncle. Identical handwriting, identical doodles. Pretty wild.

Anyhow, my uncle despised authority, and I've followed suit. I fought endlessly with my army colonel father, eventually stealing his car when I was eighteen and running off to California with a couple of college buddies. And now I despise much of the "authority" of the science world.

Maybe it's in the genes. What I do know is that I've studied the anti-science movement for about fifteen years now and can tell you that the leadership of the science world has no clue how to deal with it. The science leadership is inept—forever taking the clinical high road and not engaging. This may have been an option when I was in college, but today it is not. The whole issue is messy, but so are people.

If you want to see how fundamentally science does not understand humans, just look at the grant proposal review process for the National Science Foundation. The NSF funds research projects, and it funds communication projects. It treats them the same in terms of putting out calls for proposals, sending out proposals for peer review, and then holding panels to evaluate the proposals. I was a reviewer many times and took part in panels for both research and "informal science education."

It stunned me how the two were handled identically. Research and communication are not the same thing. Research needs aspirations of perfection. Communication is destroyed by aspiring for perfection. I could talk at great length about this, but there's no point because . . . scientists don't listen.

The bottom line is that science has to figure out how to deal with these annoying creatures called humans. There are no formulas or evidence-based strategies yet that will solve the problems. You need people on your team who possess the intuition needed to understand the social dynamics involved.

The Path Less Traveled

When I broke free of the confines of my tenured professorship in 1994, I also left the constraints of the academic science world. Most everything I've done since has paid little heed to those binds.

For example, I made a low-budget feature film in 2008 titled *Sizzle: A Global Warming Comedy*. I managed to premiere it at two major film festivals, and then we held over one hundred public screenings, everywhere from NASA to the Smithsonian Institution.

One of the most fascinating events was speaking at an event for the American Association for the Advancement of Science's Media Fellows program, where I showed a clip of the movie and talked about it. The audience members quizzed me on the source of funding—what grants did I write, what foundations funded it—and who was on my "advisory committee." I looked back at them blankly and just said, "None." Which they didn't quite get.

The answer was nobody. No one. No grants, no foundations, no advisory committee. I took $120,000 of my own money, recruited all my actor and film-maker friends in Hollywood, and just made it.

I said to the audience, "Believe it or not, it is possible to actually *do* things related to science that don't have to be run through countless channels for everyone to impose their elements of control to make sure no boats get rocked." I just did it. Which tends to strike terror into the hearts of scientists.

Some people get so deeply enmeshed in the massively organized and bureaucratic world of science that they lose the ability to think and act autonomously. That is exactly where science loses the human element—when it gets so sterile, cerebral, and controlled. This is more important than ever because, as I've said, we live in an increasingly media-driven world, yet (citing Michael Crichton again) scientists don't understand media.

Now let's take a moment to consider in a little more depth why this might be.

Why Might Scientists Not Understand Media?

I think it could well be a developmental problem. This is just a hypothesis, but I'd like you to bear with me for a moment as I lay it out, since I don't think anyone has ever suggested this.

Point 1: The brain isn't finished developing until age twenty-five. Over the past decade, neuroscience has established that the human brain does not complete its development until about age twenty-five. These new findings have changed the thinking for many programs in society. For example, juvenile justice programs have managed to use these findings to end capital punishment for juveniles in almost all states. The realization is that their brain hasn't yet finished developing; it's too soon to terminate their life rather than believe it can still be reprogrammed properly.

Point 2: Narrative stimulates the brain. A number of neurophysiologists are now working on how the brain responds to narrative. The pioneer for much of this work has been Uri Hasson at the Princeton Neuroscience Institute. I've been hounding him for five years. Last year I spent two hours in his lab telling him about the ABT as he showed me samples of his team's work. I've written about it in my last two books. The bottom line is that narratively structured material excites the brain; non-narrative material, not so much. The study of the brain and narrative is still in its infancy, but it's clear that if you tell a good story, brains get excited (duh). And the presentation of just information doesn't do this as much.

Point 3: Most scientists begin their intensive science education in their teens. Most of the professional scientists I know were already headed down the science education pathway by the time they began college. That means by their late teens they were consuming an educational diet of almost entirely science content, which is heavily informational. This means the last decade of their brain development took place in this realm.

Point 4: Science education is largely non-narrative. Even though the doing of science is a truly narrative process built entirely around the problem-solution dynamic, which is the essence of narrative, the teaching of science remains largely non-narrative. What I mean by this is your science textbooks are still largely facts, facts, facts. Whether it's biology, chemistry, physics, or math, it's delivered to you in huge swaths of fact after fact after fact. Very little effort is put into weaving it all around a central narrative. To their credit, many biology texts are built with an evolutionary perspective, following renowned geneticist Theodosius Dobzhansky's dictum that "nothing in biology makes sense except in the light of evolution." But for the most part, science is taught through facts. And, more important, how many novels do science students purchase at the bookstore every semester for science classes? Not many.

Point 5: Media is narrative. As I said in chapter 5, media is driven by narrative content. Those who communicate through the "And, And, And" mode don't find themselves on television. It's that simple.

If you add all this up, you get what I think accounts for a major part of why scientists don't understand media. The last decade of their brain development takes place in a soup of non-narrative content. Contrast this to humanities students, who are force-fed massive amounts of narrative content in the mountains of novels they are required to read, and you begin to see the basic environmental differences between the rearing of the two groups. Then add this to the large amount of research on linguistics showing that older people find it extremely difficult, if not impossible, to get rid of an accent (look at diplomat Henry Kissinger, who immigrated to the United States from Germany and never lost his thick German accent).

What I would hypothesize is that by the time a scientist is thirty-eight, his brain has gone down a developmental pathway in which neurons have formed that will never be reversed. Ever.

What makes me think this with such certainty? I know of one experiment that supports it.

A Brain Transplant Experiment

The hypothesis I'm putting forward here is new. I know of no studies to support or refute it so far, but I do know of one very relevant set of observations.

An experiment was conducted in which a young man incubated his brain in the intensely non-narrative world of science education all the way up to age thirty-eight—all the way to achieving tenure as a professor of marine biology.

Then, at the ripe old age of thirty-eight, he transplanted his brain into the intensely narrative world of Hollywood, where he has worked and resided ever since. The subject in this experiment is, of course, me.

What I can tell you is that while I've gotten a whole lot better with narrative—good enough to write three books about it—I am now willing to accept that I will never, ever be able to tell stories with the sophistication, nuance, complexity, and coherence of the great screenwriters of Hollywood.

As hard as I've tried—three years of film school, at least five writing courses, countless film projects that all told stories—I've been able to make only so much progress.

I'm sure in the end it's the usual mix of genes and environment, but I do think this is a potentially important hypothesis to consider. You cram young, developing brains full of non-narrative content, what do you expect after enough years?

Make Science Human

So that is my final message. It is so important that the world of science not blindly continue on its juggernaut of stamping out the humanities while complaining about "not enough funding for science."

The future of science will continue to be challenged by its ability to interface with society and not lose control of knowledge that can potentially go in the wrong direction. The ability of science to achieve this rests in the ability of scientists not to subsist on the periphery of society but to actually share fluently in the leading of society.

I still thoroughly enjoy working with scientists. In particular, my Story Circles Narrative Training program is producing constant breakthroughs in helping scientists see the power of narrative. As I write this, we have six Story Circles running with thirty-two research scientists with the National Park Service in Colorado.

We recently had a conference call with Larry Perez, communications coordinator for the NPS division in Fort Collins, Colorado, who is overseeing the six circles. They are on their seventh week of the ten weeks of sessions. On the call, Larry said, "I continue to be amazed—every session I sit in on produces at least one moment of awakening and realization as the five members of the circle see another aspect of how narrative works."

It is a long process, absorbing how deep and important narrative is. Actually, as Christopher Vogler said, it is a lifelong journey for even the best. In fact, in 2015 I was at a Hollywood event and ended up speaking with Eric Roth, who won the Oscar for the screenplay for the movie *Forrest Gump*.

I asked him whether, at age seventy, he felt he finally had "the narrative thing" down. He chuckled and said not by any stretch. He said that with every new screenplay, he finds himself having those same moments of awakening and realization as the Story Circles members.

Which means that the journey is endless, but it's also infinitely important. And it is the central element in making anything human. So that is my final message: continue to "be the voice of science." But also know that in being that voice, the biggest challenge of all is to understand the importance of narrative and to bring principles of narrative into your work. Only then will we achieve the ultimate goal: to make science human.

Filmmaking for Scientists

With the advent of YouTube and innovations in video technology, a new day has dawned for the communication of science. Every university I visit now has numerous science faculty members and students making their own videos. It's very inspiring and is all part of the gradual humanization of the sciences.

In our annual video-making workshops in the summer graduate course at Scripps Institution of Oceanography, we get to see students go from zero to sixty—as in most have zero experience to begin with but, over a couple of days, go to sixty seconds (they make one-minute videos).

With this in mind, here are a few simple pointers for beginners.

Work within the Constraints

Film is a very powerful medium, in both directions. You can excite an audience into actual physical action, but you can also bore your audience so badly they will forever associate your subject matter with boredom. You are welcome to adopt a "damn the torpedoes" attitude to film production—saying, "I don't care what you say; I want to pack my film with all the facts," but if you do, you'll simply be trying to pound a square peg into a round hole.

You'll do best to accept the constraints of the medium and work around them. Film is heavy on style, light on substance: the very opposite of science. Which just means you need to translate your substance into style. Here are a few ways to do that.

It's a Visual Medium

Academics lecture, but filmmakers, first and foremost, tell their stories with images. In film school, the professors didn't let us touch dialogue for the entire first year—we had to make silent movies. They wanted to show us that the visual channel is much more powerful than the audio.

The rule of thumb is that a good film can convey its basic message even with the volume off. This means that footage of "talking heads" (video of just people being interviewed) doesn't work by itself. If you turn off the volume, you can't tell whether they are talking about molecular biology or football. That's not to say that you don't want a little bit of footage of your interview subjects, only that talking heads alone are not sufficient if you're trying to reach the broad audience. Overall, a good rule of thumb is that when you finish the first cut of your film, you should show it to strangers with the volume off and see if they can get your message.

Story Structure

There's no limit to this. I talked about the basic elements in chapter 3. Believe it or not, those elements are relevant all the way down to the shortest of films. In a matter of seconds you can tell a quick, catchy story. Just look at television commercials, which can tell a whole story in less than a minute.

The Heart of a Story Is the Source of Tension or Conflict

This is the basic rule they pounded into our heads in film school, and it is endlessly true. Show me a boring documentary and I'll most likely show you a piece of work that lacks any tension or conflict.

So you say, "How do I find tension in a film about mitochondrial DNA?"

The answer lies in questions—the scientific equivalent of standard narrative conflict or tension. You tell your story in three acts. The first act begins with exposition (a brief description of the "system" in which you are working) and ends with "the inciting incident" (as they say in screenwriting), which is the formulation of a question. This becomes the glue that holds viewers and makes them want to keep watching—they're left with the feeling "I'm not turning this video off until I find out the answer to the question." I think of this every time I watch my local television weatherman, who starts his report with a trivia question, which he answers at the end. So many nights I'll have heard enough of his report but won't change the channel until the end, when he answers the darn trivia question. It really works.

In the second act, you explore possible answers to the question (hypotheses). At the end of the second act, you bring it to a climax when you reveal the key piece of information that will answer the question. In the third act, you pull the information together to answer the question, and then you wrap things up with an ending that releases all the tension and leaves the viewer with a feeling of satisfaction.

The Power of a Good Story Rests in the Details

Enough with the generalities. Last week I sat through a day of environmental talks. You know what I remember from that entire day? Only one thing—the story a guy told about how he was sitting on an airplane and the lady next to him asked for cream for her coffee, but when they brought her the small plastic containers of cream, she said, "No thanks; the plastic isn't biodegradable." And he thought to himself, "I can hardly hear her over the jet engines that are burning up fifty gazillion barrels of fuel a minute, and she's worried about a thimble-sized piece of plastic?"

That's all I remember from that day. Why is that? It's the power of a well-told story that is also very specific. Stories that are full of vague generalizations are weak. Specifics give them strength.

Arouse and Fulfill

As explored in chapter 2, much of science is so alien to the general public that it is difficult for them to connect with. The most effective way to achieve the initial arousal is often to reach into the world of humanities. Stephen Jay Gould did this wonderfully in his *Natural History* essays by leading off with tidbits from the world of sports (baseball), cartoons (Mickey Mouse), opera, architecture, painting—all of which served to grab the interest of the nonscientist, who could then be led into the world of science. The key element is to make sure you've "aroused" your audience before you begin hitting them with the facts (i.e., the fulfillment). And just take a look at most popular television commercials. They generally open with a piece of arousal (like the duck in the Aflac commercials), then segue into the fulfillment, what the commercial is about (the actual description of the product), and then offer a little entertaining tag at the end to leave you with a thought or a smile.

Casting

You're welcome to rail against the injustice of the fact that a handsome young man is more appealing to watch than a crusty old codger, but at the end of the day it's best just to accept it and work within this constraint of the medium. A viewing audience doesn't care how many Nobel Prizes your on-camera spokesman has. If he has a wildly twitching eyebrow, they aren't going to hear a word he says. Casting is tough business. Actors spend all day basically being told they aren't very attractive as they get rejected for parts. Are you tough enough to deal with this? Nobody said making *effective* videos is easy.

Show Us, Don't Tell Us

In the end, this is the most important overriding principle. Academics, being so accustomed to lecturing, end up believing in the power of words and making the mistake of thinking they can just tell their audience their message through film. If only it were that simple.

At the start of my science career, the first time I went out into nature with my thesis advisor, Ken Sebens, I came back and told him what I had seen that was so interesting. He said he didn't want to hear about it until I could *show* him what I was talking about—specifically, with data. That is what science is about. It does no good for a scientist to stand up at a meeting and tell the audience about what he thinks is happening in nature. He has to show the audience the data and then allow them to decide for themselves if that's what is really going on.

It's the same thing with filmmaking. Just as a scientist has to collect data, a filmmaker has to collect film. Which can be very tedious. But it's the same basic process. If your interview subject says the forests are dying, you have to go find film of dying forests to show your viewers in order to get them to really grasp what is being said.

All of which goes back to my initial explanation of why I've never felt that my two careers, science and filmmaking, are that different. They are both exercises in storytelling. And thus they conform to very similar rules when it comes to doing them right.

So, do your best to see the parallels and you'll find that making an effective film, in the end, is really not that different from conducting an effective scientific study.

Acknowledgments

It's 2017, and there are a few more people to thank. For all the folks I paid my gratitude to in the first edition, I'm still forever indebted. Here are some additions.

First and foremost is Todd Baldwin. He edited the first edition and was kind enough to come back and give a pass at this edition. Equally important and an equally good sport was the editor for this edition, Rebecca Bright. And a real treat from Island Press was getting to work with the same wonderful team as before—Sharis Simonian and Pat Harris.

Next in the line of key players is my Story Circles co-creator, Jayde Lovell. It was this book that inspired her, three years ago, to quit her job in Australia with a soulless public relations firm, move to New York City, dedicate herself to improving science communication, and send me a silly fan e-mail. That little message was so perfectly written that within a month I had put her to work helping me create the first prototype of the Story Circles Narrative Training program. The rest has been, and continues to be, the best professional collaboration of my life. I hope she will take the torch of troublemaking from me when the day comes that I decide to take up surf travel full-time.

Jayde has become part of a group of characters who are basically my Story Circles incubation team. We trade endless group e-mails in which we do our best to explain the ABT dynamics to one another as we continue to learn as much about this wonderful narrative template as do the Story Circles participants themselves. This group includes Park Howell of the Arizona State University School of Sustainability, Mike Strauss of the U.S. Department of Agriculture, Shirley Malcom (of the soon-to-be famous Shirley's Law), Rick Nelson of the U.S. Fish and Wildlife Service in North Dakota, Peter Griffith of NASA, Aaron Huertas, and Liz Foote.

Closer to home, I need to thank my two Malibu spiritual gurus, Laura Rosenthal and Julie Carmen. They serve as endless sounding boards for my general confusion about human behavior. Mike Backes has been my intellectual guru and idea generator for two and a half decades now. My friend Samantha has been a co-conspirator. There's a hermit crab I know who is thankworthy. And, of course, at the end of the day, my eternal appreciation for my mother, Muffy Moose. As I said in the opening dedication, she always ended her advice to me with one simple prompt: "You gotta shake 'em up!"

Notes

Introduction to the Second Edition

p. 2. C. Vogler, *The Writer's Journey: Mythic Structure for Writers*, 3rd ed. (New York: Michael Wiese Productions, 2007).

p. 2. J. Campbell, *The Hero with a Thousand Faces* (New York: Pantheon Books, 1961).

p. 7. M. Winkler, "What Makes a Hero?" (TED-Ed video, 2013).

p. 7. C. Mooney and S. Kirshenbaum, *Unscientific America: How Scientific Illiteracy Threatens Our Future* (New York: Basic Books, 2009).

p. 9. E. Yong, "What Exactly Are People Marching for When They March for Science?," *Atlantic,* March 7, 2017.

p. 12. J. Oliver, *Last Week Tonight,* HBO, Season 4, Episode 14, 4 June 2017.

pp. 13–14. M. Crichton, "Ritual Abuse, Hot Air, and Missed Opportunities," *Science* 283, no. 5407 (5 March 1999): 1461–1463.

p. 16. G. Graff and K. Birkenstein, *"They Say / I Say": The Moves That Matter in Academic Writing* (New York: W. W. Norton, 2015).

Chapter 1: Don't Be So Cerebral

p. 19. R. Olson, "Shades of Gray," letter to the editor, *Premiere,* September 2000, p. 14.

p. 22. B. Woodward, *Bush at War* (New York: Simon and Schuster, 2002).

pp. 23–24. B. Branden, *The Passion of Ayn Rand* (Garden City, NY: Doubleday, 1986), pp. 208–209.

p. 28. M. Hirschorn, "Thank You, YouTube: DIY Video Is Making Merely Professional Television Seem Stodgy, Slow, and Hopelessly Last Century," *Atlantic*, November 2006.

p. 30. Aldo Leopold Leadership Program, Woods Institute for the Environment, Stanford University, www.leopoldleadership.org.

p. 30. R. Hayes and D. Grossman, *A Scientist's Guide to Talking with the Media: Practical Advice from the Union of Concerned Scientists* (New Brunswick, NJ: Rutgers University Press, 2006).

p. 40. H. A. Orr, "Devolution: Why Intelligent Design Isn't," *New Yorker*, 30 May 2005.

pp. 44–45. J. D. Watson, *The Double Helix: A Personal Account of the Discovery of the Structure of DNA*, edited by G. S. Stent (London: Weidenfeld and Nicolson, 1968; reprint, New York: W. W. Norton, 1981).

p. 45. M. Gladwell, *Blink: The Power of Thinking without Thinking* (Boston: Little, Brown, 2005).

p. 50. National Academies of Sciences, Engineering, and Medicine, *Communicating Science Effectively: A Research Agenda* (Washington, DC: National Academies Press, 2017).

Chapter 2: Don't Be So Literal-Minded

pp. 55–56. The name "Kevin Norton" is fictitious for privacy reasons.

p. 59. K. Auletta, "The New Pitch: Do Ads Still Work?," *New Yorker*, 28 March 2005.

p. 60. D. Mattera, *Memory Is the Weapon* (Johannesburg: Ravan Press, 1987).

pp. 62–64. Pew Oceans Commission, *America's Living Oceans: Charting a Course for Sea Change* (Washington, DC: Pew Charitable Trusts, 2003). Final report available for download at www.pewtrusts.org.

p. 63. A. C. Revkin, "U.S. Is Urged to Overhaul Its Approach to Protecting Oceans," *New York Times*, 5 June 2003, p. A22.

p. 64. P. Shenon, "Sept. 11 Commission Plans a Lobbying Campaign to Push Its Recommendations," *New York Times*, 19 July 2004.

p. 65. D. Wilmot, J. K. Sterne, K. Haddow, and B. Sullivan, *Turning the Tide: Charting a Course to Improve the Effectiveness of Public Advocacy for the Oceans* (Capitola, CA: Ocean Champions, 2003). Final report available for download at www.ocean champions.org.

pp. 66–67. Numbers in figure 2-1 are from www.boxofficemojo.com.

p. 67. D. Halberstam, *The Powers That Be* (New York: Knopf, 1979).

p. 67. J. Groopman, "Being There," *New Yorker*, 3 April 2006, pp. 34–39.

p. 69. M. Kakutani, "Is Jon Stewart the Most Trusted Man in America?," *New York Times*, 15 August 2008.

p. 71. *Talking Science: The Elusive Art of the Science Talk* was a twenty-minute video I made in 1999 by interviewing faculty members from the USC departments and schools of cinema, communications, theater, biology, and physics. It's not the definitive "how to give a science talk" video, but it does hit on some interesting and important aspects of how scientists care so little about effective communication that they're willing to sit through one poor presentation after another at their science meetings. It is theoretically still available on DVD from the USC Wrigley Institute for Environmental Studies (http://wrigley.usc.edu).

pp. 74–75. L. Cuban, *Oversold and Underused: Computers in the Classroom* (Cambridge, MA: Harvard University Press, 2001).

pp. 74–75. L. Cuban, *Teachers and Machines: The Classroom Use of Technology Since 1920* (New York: Teachers College Press, 1986).

pp. 84–87. "Great Barrier Reef: Dying Down Under," *Real Sports with Bryant Gumbel*, Episode 244, produced by Chapman Downes, 18 July 2017.

pp. 84–87. *Chasing Coral*, directed by Jeff Orlowski (Netflix Productions, 2017).

Chapter 3: Don't Be Such a Poor Storyteller

p. 92. L. Grieveson and P. Kramer, eds., *The Silent Cinema Reader* (London: Routledge, 2003).

pp. 100–102. R. McKee, *Story: Substance, Structure, Style, and the Principles of Screenwriting* (New York: ReganBooks, 1997).

pp. 102–105. P. B. Medawar, "Is the Scientific Paper a Fraud?," *The Listener* 70 (12 September 1963), pp. 377–378; reprinted in *The Threat and the Glory: Reflections on Science and Scientists* (New York: HarperCollins, 1990).

p. 109. J. Hopkins and D. Sugerman, *No One Here Gets Out Alive* (New York: Plexus Publishing, 1980).

p. 109. M. Crichton, *Travels* (New York: Harper Perennial, 2002).

p. 112. R. Carson, *Silent Spring* (Cambridge, MA: Riverside Press, 1962; reprint, Boston: Houghton Mifflin, Mariner Books, 2002).

pp. 112–116. *Too Hot Not to Handle*, written by S. J. Hassol, produced by L. David, J. Glover, J. F. Lovett, L. Lennard, and M. S. Kaminsky (HBO Pictures, 2006).

pp. 112–116. *An Inconvenient Truth*, written by A. Gore and B. West, produced by L. David, L. Bender, S. Z. Burns, and L. Chilcott (Participant Productions, 2006). Danish biologist Kåre Fog provides a detailed examination of *An Inconvenient Truth* online at www.lomborg-errors.dk/Goreacknowledgederrors.htm.

pp. 114–115. W. J. Broad, "From a Rapt Audience, a Call to Cool the Hype," *New York Times*, 13 March 2007.

pp. 114–115. P. Jones, review of *Too Hot Not to Handle*, www.dvdtalk.com, 12 September 2006.

pp. 118–119. *Manufacturing Consent: Noam Chomsky and the Media,* directed by M. Achbar and P. Wintonick (Zeitgeist Films, 1992).

p. 123. R. Olson, *Houston, We Have a Narrative: Why Science Needs Story* (Chicago: University of Chicago Press, 2015).

p. 124. R. Olson, "Slow-Motion Disaster below the Waves," *Los Angeles Times,* 17 November 2002.

Chapter 4: Don't Be So Unlikeable

pp. 128–130. J. Steinbeck, *The Log from the Sea of Cortez* (New York: Penguin Books, 1951).

p. 132. *The Century of the Self,* written by A. Curtis, produced by A. Curtis, L. Kelsall, and S. Lambert (BBC Four, 2002).

p. 139. P. J. O'Rourke, "The Greenhouse Affect," originally published in *Rolling Stone;* reprinted in *The Rolling Stone Environmental Reader* (Washington, DC: Island Press, 1992).

p. 139. E. Hoffer, *The True Believer: Thoughts on the Nature of Mass Movements* (New York: Harper Perennial, 1951).

p. 145. R. A. Lanham, *The Economics of Attention: Style and Substance in the Age of Information* (Chicago: University of Chicago Press, 2006).

pp. 147–148. S. D. Levitt and S. J. Dubner, *Freakonomics: A Rogue Economist Explores the Hidden Side of Everything* (New York: William Morrow, 2005).

pp. 148–149. C. Klosterman, "The Drowsy Dozen: An Impassioned Plea for Professional Jurors, from a Man Who Just Spent a Long Time as an Amateur," *Esquire,* 31 December 2005.

p. 152. E. Castronova, *Exodus to the Virtual World: How Online Fun Is Changing Reality* (New York: Palgrave Macmillan, 2007).

p. 153. A. Gawande, "The Mistrust of Science," *New Yorker,* 10 June 2016.

pp. 155–156. N. Duarte, *Resonate: Present Visual Stories That Transform Audiences* (Hoboken, NJ: John Wiley & Sons, 2010).

Chapter 5: Don't Be Such a Poor Listener

p. 160. P. Kareiva, "If Our Messages Are to Be Heard," *Science* 327, no. 5961 (2010): 34–35.

p. 164. M. E. Mann, R. S. Bradley, and M. K. Hughes, "Northern Hemisphere Temperatures during the Past Millennium: Inferences, Uncertainties, and Limitations," *Geophysical Research Letters* 26, no. 6 (15 March 1999): 759–762.

p. 164. J. Romm, *Language Intelligence: Lessons in Persuasion from Jesus, Shakespeare, Lincoln, and Lady Gaga* (North Charleston, SC: CreateSpace Independent Publishing, 2012).

p. 165. Michael E. Mann, *The Madhouse Effect: How Climate Change Denial Is Threatening Our Planet, Destroying Our Politics, and Driving Us Crazy* (New York: Columbia University Press, 2016).

p. 165. Michael E. Mann, "I'm a Scientist Who Has Gotten Death Threats. I Fear What May Happen Under Trump," editorial, *Washington Post*, December 16, 2016.

p. 174. *The War Room*, directed by C. Hegedus and D. A. Pennebaker (October Films, 1993).

p. 175. *4 Little Girls*, directed by Spike Lee (HBO Documentary, 1997).

Chapter 6: Be the Voice of Science!

pp. 208–209. U. Hasson, O. Landesman, B. Knappmeyer, I. Vallines, N. Rubin, and D. J. Heeger, "Neurocinematics: The Neuroscience of Film," *Projections* 2, no. 1 (1 June 2008): 1–26.

Index

Page numbers followed by "f" refer to
graphics.

ABT. *See* "And, But, Therefore" narrative
    template
Academy Awards, 28–29, 48, 114, 211
acting teacher, 1–4, 20, 25, 26–27, 29, 39–40,
    54, 70, 93, 125, 158–160, 163, 187
advertising, 59, 65–67
    Hollywood movies vs. scientific report,
      65–67, 67f
    new media and, 58–60
Aflac ad campaign, 59, 216
Agassiz, Louis, 181
Aldo Leopold Leadership Program, 30
Amanpour, Christiane, 21
Ambady, Nalini, 149
American Geophysical Union meeting, 33
American Pie (movie), 34
American Samoa and educational films,
    75–76
Am I Making Myself Clear? (Dean), 204

"And, But, Therefore" narrative template, 6,
    17, 82, 85, 123-124, 125, 160, 165, 166,
    168, 169, 170–171, 174–175, 177–178,
    179, 209
Antarctica, 11, 90, 82, 108–109, 191–192
antiplot, 100-102, 101f, 194
Aquino, Valerie, 8
archplot, 100-102, 101f, 108
arouse-and-fulfill strategy, 71–74, 72, 78–79,
    81, 82–83, 91, 99, 215–216
arrogance/condescension
    avoiding, 130-131
    effects, 130-133
    examples, 130-131, 132-133
    art
    criticism and, 105-107
    transition of, 193-196, 194f
Ashley, Courtney, 127, 129
Atlas Shrugged (Rand), 23
Auletta, Ken, 59
Australian Institute of Marine Science, 91
Aykroyd, Dan, 33

Back to the Future (movie), 137

Baldwin, Alec, 159-160

Barnacles Tell No Lies (film/Olson), 44, 80-81

Baron, Nancy, 204

"Being There" (Groopman), 67

Belushi, John, 33

Benigni, Roberto, 29

Bernays, Edward, 132

Big Brother (TV show), 28

Biographia Literaria (Coleridge), 99

Black, Jack, 150, 150f, 151, 181

Blink (Gladwell), 45, 149

Blogosphere, 140-142

    anger and, 140, 141-142

Bogart, Humphrey, 92-93

Bolin, Ralph, 129

boredom definition, 110

boredom "errors"

    accuracy errors vs., 110-117, 115f, 116f

    examples, 89-91

Branden, Barbara, 23-24

Branden, Nathaniel, 23-24

Brennen, Tim, 34

Brosnan, Pierce, 179

Bryan, William Jennings, 62, 148

Buñuel, Luis, 80

Bush, George W./administration, 22, 114

Calvert, John, 96

Cameron, James, 122

Carlisle, Ty, 95-97

Carson, Johnny, 100, 183, 203

Carson, Rachel, 112

Cartesian coordinates, 44

Carville, James, 169-172, 172f, 173-174

Casablanca (movie), 92-93

Castronova, Edward, 152

Century of the Self, The (documentary/ Curtis), 132

Chain, Ernst, 31

Chan, Jackie, 100

Chapman, Matthew, 186

Cheers (TV show), 28

Chicago (movie), 94

Chomsky, Noam, 119

Christo, 195

Citizens for Tax Reform, 65

Clancy, Tom, 192

Clinton, Bill, 22, 148, 169, 171, 174

Clinton, Hillary, 155, 164, 167–168, 168f, 169–171, 173–176

Cluetrain Manifesto, The: The End of Business as Usual, 197-198

Coleridge, Samuel Taylor, 99

Comedy Central, 34, 140

communication

    arouse-and-fulfill strategy, 69-71, 72, 79, 81, 91, 99, 146

    positive feedback loop/snowball effect, 189-190

    See also science communication; specific types

concision, 119-121

condescension. *See* arrogance/ condescension

Coolidge, Jennifer, 34

Coolidge, Martha, 199

Cosmos: A Personal Voyage (TV series), 182

Costello, Jimmy, 128

Cotton, F. Albert, 184

coupons in cereal box story, 117-118

creativity/discipline spectrum, 137-138, 138f

Crichton, Michael, 13-14, 17, 109, 160-163, 176, 192, 208

Crick, Francis, 44-45,45f

critical thinking, 61, 134-136, 143

Cuban, Larry, 74-75

Curb Your Enthusiasm (HBO series), 113

cynicism in science, 136, 138, 138f, 139, 186-187, 201

Dada cultural movement, 193-195

Daily Show with Jon Stewart, The (TV show), 69

Darrow, Clarence, 62

Darwin, Charles, 30-31, 186

David, Larry, 113

David, Laurie, 11, 113, 115, 116f, 195

David and Lucile Packard Foundation, 65
Dean, Cornelia, 204
Descartes, 44
"Devolution: Why Intelligent Design Isn't" (Orr), 40, 95
discipline/creativity spectrum, 137-138, 138f
Discovery Institute, 96, 98
Do the Right Thing (movie), 89
Doors, 109
Double Helix, The (Watson and Crick), 44-45
Dragons of Eden, The (Sagan), 182, 183
"Drowsy Dozen, The" (Klosterman), 149
Druyan, Ann, 203
Dubner, Stephen, 147-148
Duchamp, Marcel, 194, 194f, 195
"dumbing down" vs. concision, 120-121

East Coast Benthic Ecology Meetings (1978), 145
Economics of Attention, The: Style and Substance in the Age of Information (Lanham), 145, 146, 148
Edison, Thomas, 74-75, 78
Eisenstein, Sergei, 80
"elevator pitch," 119-120
Ensler, Jason, 19, 127
Escape from the Ivory Tower (Baron), 204
Esquire magazine, 148-149
Evita (movie), 94
evolution vs. intelligent design debate
    about, 40, 41, 46, 60-61, 94, 118, 131-132
    See also Flock of Dodos: The Evolution-Intelligent Design Circus (film/Olson)
Exodus to the Virtual World (Castronova), 152

"false negative"/"false positive," 111-112
Farley, Chris, 33
Ferrell, Will, 34
Field, Sally, 29
films
    audience/film as visual, 77-78, 214, 215-217

casting, 216
    as educational medium, 73-76, 78-79
    juxtaposition and, 79-81
    as motivational, 76
    scientists finding inaccuracies in, 19, 100, 110, 121-122, 190
    scientists making, 213-217
    tips for scientists making, 213-217
Fleming, Alexander, 31
Flock of Dodos: The Evolution-Intelligent Design Circus (film/Olson)
    about, 40-41, 46, 54, 79, 116, 192
    arrogance/condescension and, 132
    finding the story, 95-99, 97f, 109
    screenings/responses, 4, 59, 60, 105, 112, 120, 144, 203
Fog, Kåre, 114
Fountain (Duchamp), 194, 194f, 195
Four Horsemen of the Apocalypse, The (movie), 92
four organs of mass communication theory
    about/examples, 20-24, 21f, 39-40, 46, 61, 73, 76, 106, 152, 155, 181
    See also specific organs
Freakonomics (Levitt and Dubner), 147-148
Freud, Sigmund, 43, 132

Gaither, Daniele, 34
Gawande, Atul, 153
Gipsy Kings, 34
Gladwell, Malcolm, 45, 149
Goldwater, Barry, 80
Gore, Al
    Inconvenient Truth, An and, 12, 83, 112-116, 115f, 116f, 162, 195-196
Gould, Stephen Jay, 72, 91, 105, 137, 142, 216
Great Barrier Reef, 72, 80, 85, 154, 191
"Greenhouse Affect, The" (O'Rourke), 139
Greenspan, Alan, 24
Grisham, John, 192
Groopman, Jerome, 67
Groundhog Day (movie), 75
Groundlings Improv Comedy Theater/cast, 34, 36, 47, 48-49, 57, 126, 179

Guggenheim, Davis, 115
gut (mass communication theory)
    overview, 9f, 10, 20, 21f, 22, 23f, 25, 43-46
    See also intuition

Halberstam, David, 67
Hansen, James, 114, 115
Hartman, Phil, 34
Harvard University alumni trips, 90
Hawking, Stephen, 146
head (mass communication theory)
    overview, 20-24, 21f, 23f, 25
heart (mass communication theory)
    overview, 20-24, 21f, 25, 61
"high concept," 120
Hoffer, Eric, 139
Hollihan, Tom, 71
Hollywood and mass communication, 6,
    13-14, 17-18
Hopkins, Jerry, 109
Houston, We Have a Narrative (Olson), 123,
    125, 166

Ice Blue (Olson), 191-192
Iconoclasts (TV show), 21
improv acting/training
    about, 29, 33-35, 37, 39
    "add-on story game," 36-39
    Groundlings Improv Comedy Theater/
        cast, 34, 36, 47, 48-49, 57, 126, 179
    use examples, 36-37, 38, 39, 41, 43, 47-48
Inconvenient Truth, An (movie)
    about, 112-113, 176
    accuracy/boredom errors and, 112-116,
        115f, 116f
    significance of, 112, 115-117, 195-196
Ingram, Rex, 92
Inherit the Wind (movie), 62
intelligent design. *See* evolution vs.
    intelligent design debate
Internet Movie Database, 115, 115f
interviews
    improv approach, 37, 38, 39, 43
    recommendations on, 143-144
    science approach, 37-38, 41-43, 47-49

intuition
    about, 22, 23f, 26-27, 39-41, 43-46
    definition, 44
    science and, 43-44, 45-46, 45f, 49-51
"Is the Scientific Paper a Fraud" (Medawar),
    103-105

Jacks, Samantha, 179–180
Jackson, Janet, 43, 56
Jackson, Jeremy, 11, 55, 124
Jolie, Angelina, 24
Jung, Carl, 20
Jurassic Park (movie), 13, 76
juxtaposition and films, 79-81

Kattan, Chris, 34
Kekulé, 44
Kennedy, John F., 29
Kenney, Justin, 63
Kerry, John F., 22
Kirshenbaum, Sheril, 7
Klosterman, Chuck, 149
Kohn, Alan, 70
Krieger, Robby, 109

Lanham, Richard, 145, 146, 148, 149, 195
Last Week Tonight with John Oliver (TV
    show), 11–12
Lee, Spike, 89, 119, 175
legal system
    "false positive" error, 111, 112
    jurors/lawyers and likeability, 148-149
Leister, Pascal, 54
Less Than One campaign, 24-25, 151
Levitt, Steven, 147-148
Lewis Black's Root of All Evil (TV show), 140
Liem, Karel, 182
likeable/unlikeable traits
    about, 130-133, 152
    examples showing importance, 128-130,
        147-151, 150f
    "Hollywood makeover" and, 144
    See also negativity
literal mindedness of scientists/delusions

instant enlightenment/"science-think," 62-64

instant messaging, 55-60

instant victory, 57-60

Lizard Island, Great Barrier Reef, 154, 200

Lloyd, Christopher, 137

Lobstahs (film/Olson), 187

Log from the Sea of Cortez, The (Steinbeck), 128

Lomborg, Bjørn, 114

Los Angeles Times, 36, 49, 56, 124

Lovitz, Jon, 34

lower/sex organs (mass communication theory)

  about, 21f, 22-24, 26-27, 29, 39-40, 46, 71, 73, 81, 152, 155-156

  warnings, 44

"lowered standards," 57

Lowi, Jay, 127

"mad scientist stereotype," 137

MADtv, 34

Malcom, Shirley, 174–176, 176f

Mann, Michael, 163–166

Manufacturing Consent: Noam Chomsky and the Media (documentary), 119-120

March for Science (2017), 7, 8-9, 11, 16f, 18, 49, 176f

Margulis, Lynn, 185

Marshall, Richard J., Jr., 205–207, 206f

Mattera, Don, 60

McCartney, Paul, 33

McKee, Robert, 100-101, 101f

McKinley, William, 148

Medawar, Sir Peter Brian (P.B.), 103-105

Meisner technique, 29, 40, 95, 159, 199

Memory Is the Weapon/Memoirs of an Apartheid Protester (Mattera), 60

Mendel, Gregor, 30-31

Meshes of the Afternoon (movie), 101

Miller, Stanley, 184

Miller, Steven, 61, 105, 106

miniplot, 100-101, 101f,

"modern synthesis," 31

Monty Python and the Holy Grail (movie), 101

Mooney, Chris, 7, 204

Moore, Michael, 195

Morrison, Jim, 109

Moulin Rouge! (movie), 94

Mr. Wizard Science Clubs, 151

Murray, Bill, 104

My Kid Could Paint That (documentary), 195

Myers-Briggs Type Indicator, 20, 43

Napoleon Dynamite (movie), 63-64, 64f

Narrative Index, 166–167, 168, 168f, 169

National Academy of Sciences

  Carl Sagan and, 184-185, 186, 187, 203

  Science and Entertainment Exchange program, 121

National Center for Science Education, 61

National Geographic Society, 184

National Oceanic and Atmospheric Administration, 68

National Rifle Association, 65

National Science Foundation, 138, 185, 189, 207

Natural History magazine, 73, 91, 216

Natural Resources Defense Council, 113

Nature journal, 198

"Naturejobs" Website, 198

Nealon, Kevin, 34

negativity

  acting class and, 140-141

  blogosphere, 140-142

  science and, 133-138, 138f, 139, 142-143, 144-145, 186-187

"New Pitch, The: Do Ads Still Work?" (Auletta), 59

New York Times, 36, 63, 64, 69, 114, 116, 192,

New Yorker, 40, 59, 67, 95, 153

Newsweek magazine, 63, 118

Newton, 44

9/11 Commission report, 64, 66

Nissan car commercials, 58

Njoku, Ifeanyi, 57

No One Here Gets Out Alive (Sugerman and
   Hopkins), 109
Nobel Prize, 45, 112, 187, 216,
Norton, Kevin, 55-56, 58
nuclear power example, 54-55

"objectivist" school of thought (Rand), 103
observer effect, 43-44
ocean conservation
   objective vs. subjective elements, 65-66
   See also specific projects
Ocean Symphony public service announce-
   ment (Olson), 150-151, 150f, 181
octopus story, 68-69
Old Man of the Sea, 129
Oliver, John, 11–12
Olson, Randy
   background summary, 1-2, 3-4, 93-95
   title of *Don't Be Such A Scientist*, 200-201
   See also specific works
one-third rule, 185-186
Oppenheimer, Michael, 116
Oprah Winfrey Show, The (TV show), 138
Oreskes, Naomi, 12, 204
organs of mass communication. *See* four
   organs of mass communication
   theory
Origin of Species (Darwin), 30-31, 140
O'Rourke, P.J., 139
Orr, H. Allen, 40, 95
Oswalt, Patton, 140
Oversold and Underused (Cuban), 74

Palance, Jack, 29
Panetta, Leon, 63
Paris, Texas (movie), 101
Parker, Trey, 166
Passion of Ayn Rand, The (Branden), 23
Pauling, Linus, 44-45, 45f
Pauly, Daniel, 10, 55
penicillin, 31
Perfect Storm, The (movie), 19
Pew Oceans Commission study/report
   impact expectations, 56-57, 62-63, 64

marketing and, 62-63, 66-67, 67f
"pilot season," 27
Pitt, Brad, 24
plot types, 100-102, 101f
political conventions, 29
political elections and likeability, 147-148
Porter, Jim, 145
positivity/affirmation
   about, 138, 142, 143
   improv acting/training, 34-35, 36-39
   interviews and, 143-144
Poundstone, William, 182-183, 185
Powers That Be, The (Halberstam), 67
Premiere magazine, 19, 190

Radner, Gilda, 33
Rambo (movie/character), 101, 104
Rand, Ayn, 23, 103
"Randy, the" expression, 127, 129
Reader's Digest and critical thinking, 135
Reagan, Ronald, 148
reality shows, 27-28, 108
Real Sports with Bryant Gumbel (TV show),
   84–85, 86
Real Time with Bill Maher (TV show),
   164–165
religion and heart (mass communication
   theory), 21, 24, 61
Reno 911 (TV show), 34
Reubens, Paul, 34
Ricketts, Ed "Doc," 128
Rime of the Ancient Mariner, The
   (Coleridge), 99
Roger and Me (documentary/Moore), 195
Rolling Stone, 139
Romm, Joe, 164
Rosenthal, James, 149
Rotten Jellyfish Awards (film/Olson), 34
Rowley, Jeremy, 34, 36-37, 48, 57-58
Rudolph, Maya, 34
Rufe, Roger, 63
Rule of Replacing, The, 125, 166
Rumble in the Bronx (movie), 100
Run Lola Run (movie), 75

Sagan, Carl
  National Academy of Sciences and, 184-
    185, 186, 187, 203
  science communication, 100, 182-184,
    183f, 203
Saturday Night Live (TV show), 33-34
Schneider, Stephen, 166
science
  definition, 110
  descriptive to experimental transition,
    193-194, 195-196
  dual nature of, 30-33, 33f
  fraud/outcome, 110, 116-117
  humanization of, 197-198, 213
science communication
  broad vs. academic audience, 201-202,
    202f
  criticism of popular style, 184-187
  example from 1850s, 181-182
  objective/subjective divide, 32-33, 33f, 35
  opposing change, 144-145
  prearoused audience and, 72-73, 82-83,
    119, 202f
  quality, 29-30, 70-71, 214
  recommendations on, 199, 201-204
  science attackers and, 118
  scientific talks, 70-71, 144-145, 179-180,
    215
  scientists' delusions on, 55-64
  slide technique changes, 144-145
  style/substance divide, 32-33, 33f, 35-36,
    132-133, 145-147, 214
  transition after World War II, 204
  views on importance, 189
science communication/papers
  about, 103-104, 197
  critical reviewing, 105, 106, 133-134
  as "fraud," 102-105
"Science Debate 2008," 186
Science journal, 198
"science-think," 62-64
scientific method, 142-143
*Scientist's Guide to Talking with the Media,
    A* (Union of Concerned Scientists), 30,
    36, 41-42

"Scopes Monkey Trial," 61-62
Scripps Institution of Oceanography
    communication training
  evaluations, 186
  improv training, 36-37, 41
  journalists, 36-37, 41
  making videos and, 213
  script learning/forgetting, 39-40
sea serpents, 129
Sea Studios, 183
Sebens, Ken, 216
Second City, 33, 47
Seinfeld (TV show), 113
sex organs. *See* lower/sex organs (mass
    communication theory)
Shakespeare, 40
Shape of Life, The (documentary series),
    184
Shelley, Mark, 183
Shepard, Alan, 29
Shifting Baselines Ocean Media Project
    (Olson), 11, 34, 40, 56, 57, 61, 62, 178
"shifting baselines" term, 10, 55-56, 57,
    58-59, 82, 124, 125
Shirley's Law, 174–176
Showtime, 123
Silent Spring (Carson), 112
Sizzle: A Global Warming Comedy (film/
    Olson)
  about, 57, 71, 99-100, 192, 196, 204, 208
  reviews/responses, 99-100, 185-186
Society of Independent Artists, 194
South Park (TV show), 125–126, 166
spontaneity overview, 26-27, 29, 33-35
Star Trek (TV series/movies), 32
Star Wars (movie), 66, 100, 101
Steinbeck, John, 128-129, 187
Sterne, Jack, 65
Stevenson, Adlai, 67
Stewart, Jon, 69
Stone, Matthew, 125–126, 166
Story (McKee), 100-101, 101f
Story Circles Narrative Training, 7, 84, 85,
    124, 126, 176, 205, 211, 212

storytelling
  about, 90, 110-117, 111, 115f, 116f, 122,
    146
  objective/subjective parts, 92-93
  reality vs. fantasy, 99-100, 107-109
  scientist examples, 91-92, 121-123
  set stories, 109
  story structure, 92, 93, 100-102, 101f, 122,
    214
  tension/conflict and, 118-119, 214-215
Stripes (movie), 104
style/substance divide
  educational films, 76-77, 79
  excessive information and, 148-149
  science communication, 30, 32-33, 33f,
    35-36, 132-133, 145-146, 214
"subliminal seduction," 78
substance. *See* style/substance divide
Sugerman, Danny, 109
Survivor (TV show), 28
"suspension of disbelief," 99-100
Swayze, Patrick, 199

*Talking Science: The Elusive Art of the Science
  Talk* (film/Olson), 32, 71
*Teachers and Machines: The Classroom Use
  of Technology Since 1920* (Cuban),
  74-75
Tender Mercies (movie), 101
They Say/I Say (Graff and Birkenstein),
  16–17
Thomas, Alex, 58
Three Wishes (movie), 199
Time magazine, 63, 118
Tiny Fish Public Service Announcement
  (Olson), 34
Titanic (movie), 121, 122, 123, 193
title of project, 56, 57-58, 59-60
Tonight Show, The (TV show), 100, 183, 203
Too Hot Not to Handle (HBO film), 113-114,
  115-116, 115f, 116f
Top Gun (movie), 76
Travels (Crichton), 109
Tribeca Film Festival, 59, 112

True Believer, The (Hoffer), 139
Trump, Donald, 12, 125, 164, 167, 168f, 169,
  171, 174, 176
*Turning the Tide: Charting a Course to
  Improve the Effectiveness of Public
  Advocacy for the Oceans* (Wilmot and
  Sterne), 65
Turow, Scott, 192
"type one"/"type two" errors, 110, 111-112
Tyson, Neil deGrasse, 121-122, 123

undersea habitat (Hydrolab Undersea
  Habitat), 68, 108
Union of Concerned Scientists, 30, 36,
  41-42
unlikeable traits. *See* likeable/unlikeable
  traits
*Unscientific America: How Scientific Illiteracy
  Threatens Our Future* (Mooney and
  Kirshenbaum), 204
USC School of Cinematic Arts, 70, 86, 107-
  108, 107f, 191,

voice
  personal voice, 196-199
  of scientists, 19, 190-191, 198
Vogler, Christopher, 2, 6, 15, 211

Wahlberg, Mark ("Marky Mark"), 19, 190
Warhol, Andy, 195
Washington Post, The, 165
Watch Mr. Wizard (TV show), 151
Watson, James, 44-45, 45f
Wiig, Kristen, 34, 48
William Morris agency, 191-192
Wilmot, David, 65
Winkler, Matthew, 7
Writer's Journey, The (Vogler), 2, 6, 15, 211
Woodward, Bob, 22

Yarbrough, Cedric, 34
YouTube, 28, 48, 124, 213

Zellweger, Renée, 21